**e-shock**

*Also by Michael de Kare-Silver*

**STRATEGY IN CRISIS**

# e-shock

## The electronic shopping revolution: strategies for retailers and manufacturers

Michael de Kare-Silver

© Michael de Kare-Silver 1998

Softcover reprint of the hardcover 1st edition 1998

All rights reserved. No reproduction, copy or transmission of
this publication may be made without written permission.

No paragraph of this publication may be reproduced, copied or
transmitted save with written permission or in accordance with
the provisions of the Copyright, Designs and Patents Act 1988,
or under the terms of any licence permitting limited copying
issued by the Copyright Licensing Agency, 90 Tottenham Court
Road, London W1P 9HE.

Any person who does any unauthorised act in relation to this
publication may be liable to criminal prosecution and civil
claims for damages.

The author has asserted his right to be identified as the
author of this work in accordance with the Copyright,
Designs and Patents Act 1988.

First published 1998 by
MACMILLAN PRESS LTD
Houndmills, Basingstoke, Hampshire RG21 6XS
and London
Companies and representatives
throughout the world

ISBN 978-1-349-14776-2     ISBN 978-1-349-14774-8 (eBook)
DOI 10.1007/978-1-349-14774-8

A catalogue record for this book is available
from the British Library.

This book is printed on paper suitable for recycling and
made from fully managed and sustained forest sources.

10  9   8   7   6   5   4   3
07  06  05  04  03  02  01  00  99

Copy-edited and typeset by Povey–Edmondson
Tavistock and Rochdale, England

# Contents

*Preface and acknowledgements* vii

*Introduction* 1

1  Overview: the key conclusions 7
2  Definitions and consumer trends 25
3  Learning from the pioneers in electronic selling 31
4  Future growth of electronic shopping: look out for 2005! 47
5  The ES test: how to tell how much your business is going to be affected 71
6  How can retailers respond? 89
7  Ten strategic options for retailers 103
8  The store of the future 121
9  Rapidly improving technology meets growing consumer demand 131
10 The world is changing: assets to knowledge 153
11 Structural difficulties with the Internet will be overcome 161
12 How can manufacturers respond? 173
13 Manufacturers' ten strategic options 185
14 The new marketing imperatives: marketing in the electronic age 205
15 Setting the strategy and mobilising the organisation 227

*Appendix 1: The retailer dilemma – can shops still be profitable?* 245

*Appendix 2: Retail banking case study* 254

*List of figures, tables and plates* 261

*Bibliography* 264

*Index* 269

# Preface and Acknowledgements

The stimulus for writing this book came when I read announcements by Wal-Mart, Home Depot, Tesco and others that they were opening new stores, yet at the same time relaunching their web sites. Was this not a contradiction in terms? Could retailers operate both physically and electronically without one cannibalising the other? Would not electronic sales (presumably mostly over the Internet) significantly reduce numbers visiting the stores and so the profitability of the physical space? Could retailers just carry on as before investing heavily in land and buildings while the virtual world beckoned ever more eagerly? Would that new route to market eventually dominate? How many consumers were already willing and interested in shopping electronically? When would it all reach critical mass?

The purpose of this book is to answer those questions. The aim is to provide a practical guide to the future consumer retailing scene, how electronic commerce will develop, what impacts it will have, pin down when it will happen and suggest a framework to help retailers and manufacturer suppliers determine their optimum response. I have drawn on many years of consulting experience with companies in different sectors, observing and advising on market, customer and channel challenges. Other research and interviews have been carried out, tests and trials observed and evaluated and ideas checked with client companies whose confidentiality I must respect but whose ambitions and intentions are clear.

Forecasting future market developments can never be an exact science but I have also been enormously encouraged in my conclusions and convictions by a number of people. First, thanks to my partners at the Kalchas consulting group and to new colleagues at CSC. They have helped create the experiences and opportunities to write this book and have been stimulating in their own thinking and perspectives. Others have acted as important sounding boards, reinforcing some basic premises but steering me clear of unwarranted prejudices. Among this group I would like to thank David Stoddart especially, retail analyst at Henderson Crosthwaite, and intellectual soulmate, Simon Wright of Virgin, Mike Nevin of Dixons, Chris Warmoth of Procter & Gamble, a friend and deep thinker, and Svea Kordt and Ragnar Nilsson of Karstadt. A few other people have been more generally inspirational in encouraging me to put pen to paper. I am grateful to Robert Heller, whose own writings and teachings constantly challenge the status quo, Lord Peter Walker, whose own achievements continue to set a benchmark, and Walter Goldsmith, a constant fount of energy and ideas.

As for the composition of the book, it could not have happened without the vision and support of Stephen Rutt, my editor at Macmillan, and also Elaine Howells and Louise Crawford, whose marketing enthusiasm has been infectious. David Oates, author and writer in his own right, provided a critical role early on helping to organise initial musings and ideas, writing up some early drafts and providing additional context. As for the typescript, especial thanks to Jo Mountford and Lorraine Oliver who in their personal time painstakingly transcribed my physical labours into ordered electronic prose. Finally, but most importantly, a special thanks to my wife, Deborah, whose sound commercial instincts and critical appraisal have helped keep the ideas here firmly rooted in what is practical and achievable.

Michael de Kare-Silver

*Note*: The author and publishers are grateful to Netscape Communications Corporation for permission to use material on the jacket of the book. However, Netscape Communications Corporation has not authorised, sponsored, or endorsed, or approved this publication and is not responsible for its content. Netscape and the Netscape Communications Corporate Logos are trademarks and trade names of Netscape Communications Corporation. All other product names and/or logos are trademarks of their respective owners.

# Introduction

There's a new milestone being reached in business and commerce. The beginning of the twenty-first century is going to see a radical transformation in the retailing of goods and services. There's a revolution taking shape in how people shop. In just a few years it's going to reach a significant point in its evolution. Happening with remarkable speed, it's forecast to reach critical mass and start having a serious impact by around 2005. It's the advent of mass electronic shopping.

New technologies such as the Internet, Web TV, interactive kiosks and videophones are making visits to the shops increasingly unnecessary. As they develop they will enable more and more people to order from their home or their office or any place they can make an electronic connection. By just the year 2000 the Internet is predicted to reach into some 40% of households across the USA and into 20% or more homes in the UK and Germany, with other developed nations racing to catch up.

Electronic access is becoming more widespread, the equipment more user-friendly, quicker and cheaper to buy. It offers to save time and promises lower prices. It provides the convenience of 24-hour shopping to any place across the globe whenever the consumer wants it. There's no restriction on physical location, drive times, catchment areas and planning restrictions. It operates in the freedom of cyberspace.

It's new and it's exciting. It's going to be the next wave of electrical gadgetry to reach into the home. It will be volume merchandised for mass consumption. It will have the marketing push of the likes of Sony, Dell and Microsoft. Word of mouth will communicate the advantages. The media will avidly report on progress.

The latest innovations will be widely reviewed, heightening awareness, expectations and desire. Computer, telecommunication and media companies will continue to pour in massive investment. They will be looking to constantly improve and accelerate the commercialisation of electronic interactions. There's an unstoppable momentum that's gathering pace, with a new advance to report almost every week. The electronic age is drawing ever nearer and consumer shopping will never be the same again.

Gone will be the days of pushing heavy shopping carts up and down the aisles, loading up with the weekly groceries, dragging tired and reluctant children behind, waiting in line at the check-out, unloading into the trunk of the car, driving home amid heavy traffic, unloading from the car again and reloading into kitchen cupboards or the fridge freezer. The shopping chores will all be done electronically. The standard weekly food and household order will be set up with a local home delivery company. An e-mail, a phone call or fax quoting a reference number will automatically trigger a reorder to be delivered at a pre-agreed time. Specials for the week will be added by ticking off the electronic order form. Payment will automatically be deducted from the bank account through a secure encrypted payment network. The frustrations of shopping for staple and basic items will be a thing of the past.

The convenience of electronic shopping will not just be confined to food and household products. It will impact many other sectors. On the Internet today computers, books, music CDs, flowers and wine are already significant sellers. QVC has demonstrated the possibilities of selling jewellery, hardware, DIY, electricals and health and beauty products through its TV home shopping network. Catalog companies like Great Universal, Next, Spiegel and Eddie Bauer have shown that where it's well merchandised and marketed consumers will confidently buy clothes by mail order phone and fax – even before trying them on. In the financial services arena companies like Direct Line in insurance and Citibank and First Direct in consumer banking have pioneered the selling of their services electronically. ATMs have proven consumer comfort with simple easy-to-use electronic technology. No need to visit the bank branch any more when all basic banking can be done through a hole in the wall and increasingly now by phone or via the PC.

With chores out of the way, basic shopping done and delivered, financial affairs taken care of, a whole new world of consumer opportunity opens up. Suddenly there's more time for family and friends. There's the opportunity to relax, pursue hobbies and have more quality leisure time. When people do go out shopping, expectations of a more enjoyable experience will be high. Their emphasis will be on seeing new things, trying new products. Shopping will develop as more of an integrated social and leisure experience. Consumers will treat it as more of a day out. Visits to large shopping centres will grow in popularity. It will present the opportunity to see a complete range of products with all the leading suppliers and distributors represented. There will be special areas and crèches for leaving young children with highly trained staff in superb facilities. There will be a wide choice of places to eat, meet friends and socialise. In the bigger centres and malls there will be additional social and entertainment facilities – for example, with sports and fitness clubs and movie theatres.

This type of large-scale shopping environment is already seen in locations across the USA and in a few massive centres in Europe such as the Metro Centre in England and CentrO in Germany. In these places often a theme park atmosphere is deliberately created to encourage people to visit, spend an extended amount of time in the one location and explore the vast array of shops. At CentrO, for example, there is everything from the Gap, Coca-Cola, Original Levi's Store, jazz clubs, a Warner movieplex, Planet Hollywood, an Irish pub through to clothing department stores, designer fashion shops and food supermarkets. At new shopping centres being established in the USA some of the sites are being described as 'Centre Parcs meet Ikea' or 'Disneyworld meets Wal-Mart'. These sites can be so vast that they can incorporate ski slopes and canoe runs. Not surprisingly where they are established they are seeing sales booming as they attract the greater share of consumer spending in a wider and wider catchment area.

More centres are being planned in Europe. The UK alone, despite tight planning controls, will see a significant increase in major shopping developments over the next five years. There will be four new sites opening in Manchester, Dartford, Glasgow and Bristol. There will be substantial extensions to existing centres in Birmingham and Newcastle. Planning permission has already been given for two greenfield sites in Paddington and White City in west London. These will all create opportunities for retailers to start developing their 'stores of the future'.

The shopping scene is changing and retailers will need to develop. Standing still carries a high risk of being disintermediated, cut out of the supply chain as electronic ordering grows. But the changing environment is one of opportunity as well as challenge. There's a potential to forge new sources of advantage and competitive differentiation and in principle retailers remain best placed to pursue that. After all, they are closest to the end consumer, they have that relationship right now. They have built strong consumer franchises reinforced by their own brands and reward and loyalty bonuses. They have the best database of who their customers are, what are their habits and preferences.

As they move into the next century retailers will have a range of options. At one extreme they could transition their business to become a full electronic home-delivery operation gradually moving out of their physical retail estate. As an alternative they could look to revitalise their physical presence and evolve the store proposition to meet some of the changing consumer demand. Many have already begun down this path. UK grocers like Tesco and Sainsbury have expanded their service range to include banking, dry-cleaning, flowers, newspaper and magazines, coffee shops and gas stations. Others like Levi's, Doc Marten's shoes and clothing and Nike have filled their stores with video walls, museum artefacts and related merchandise to give more reasons to visit and enhance the total shopping experience.

The days of extensive shelving, racks of products, shopping carts, aisles and check-out counters will soon be over. To remain attractive retailers will have to go back to the drawing board and rethink how to use and exploit the total physical space at their disposal. They can start to experiment with totally new ways of shopping – make the experience, for example, part electronic part physical, partner with others to provide integrated social, leisure and entertainment facilities and excite consumers so they still want to visit despite the temptations of electronic shopping.

Yet as we approach the end of the 1990s the retailing community at large is still ambivalent about electronic commerce, its potential and how it will develop. They naturally fear their sizeable investments in their physical retail estates being undermined. They are reluctant to encourage their consumers to think and act electronically, not wanting to see cannibalisation of store sales and drop in store traffic.

While they hesitate in their response they are leaving themselves vulnerable in their market places. Dedicated electronic sellers like Amazon.com, Peapod, Dell, 1-800-FLOWERS, Direct Line and others have established themselves. They are unencumbered by existing infrastructures. They have no distribution channel conflicts. They can take advantage of the Internet's immediate global reach and with effective marketing quickly exploit that growing group of consumers who unequivocally prefer to shop electronically. If existing retailers don't watch their markets carefully they might find a whole segment of their customer base moving away from them, determined to avail themselves of a rival's service offering better suited to their needs. In some sectors like books in the USA, personal insurance in the UK, music CDs, flowers and computers the entry barriers are already being erected, making it more costly and difficult for an existing physical retailer to move into electronic selling – there's already entrenched electronic competition!

And as retailers experiment and remain uncertain about their strategy for electronic shopping so there is a threat from another quarter. Manufacturer suppliers, long squeezed by their consolidating and sophisticated retailer customers, now find themselves with a chance to fight back and establish their own direct links with the end consumers of their products. They can simply bypass the retail chains. And for sure they are evaluating their options. Unilever, Heinz, Procter & Gamble, Campbell and Nestlé – five of the world's biggest consumer goods companies – are all exploring the electronic world. To date their own initiatives have been cautious, not wanting to disturb their existing retail customer base until they are more certain of the size of the electronic shopping opportunity. But, for example, the day may be not far off when we see P&G, Unilever, L'Oréal, Kimberley Clark and Rubbermaid combine with a joint venture to provide 'The Household Goods Shop.com'. This could be an Internet site offering a full range of all a consumer's household goods needs, regular orders with home delivery at lower prices direct from the manufacturer.

Electronic Shock? Every part of the retail scene is going to go through a significant transformation in the coming years. Those who stand still will surely decline and risk becoming some of the early dinosaurs of the twenty first century. Those who win will be companies who develop a close understanding of what their customers will want, what will be their needs, whether they want electronic facilities for the sort of goods and services the company offers, how quickly they'll want to change and whether they'll all move that way or just a proportion. By developing this understanding of the electronic marketplace retailers *and* manufacturers will be able to develop clear long-term plans for responding to the new challenges and keeping their customers. It's all about to take place so rapidly – and reach critical mass within the next ten years – that there's certainly no room for complacency and little time left for continued ambivalence. Only by facing up to the electronic shopping revolution can future business be secured.

# 1 Overview: The Key Conclusions

It's all happening very quickly. In just five years since it went commercial the Internet is proving to be one of the biggest phenomena business has seen. Bursting onto the scene in 1994 it has caused a major rethink on how to sell products and services to consumers. Many companies are rushing to explore this new opportunity. What potential does it hold, will it impact and if so, when?

After the initial excitement there has been an inevitable period of trial and experimentation, a gradual awakening and appreciation of the realities of electronic commerce. It can provide many advantages to the consumer, it can be a new growth vehicle, it can offer a lower cost higher margin route to market. But it's going to take significant investment, there will be structural difficulties like payment security to be overcome and despite the hype existing technology is not yet up to the communication challenge. But how long till it will be?

The coming era of electronic commerce is in the early stages of growth. Like the VCR, the computer, the microwave and satellite TV there is a necessary period of acquaintanceship and improvement before it really takes off. This essential development is now taking place. The end of the twentieth century is seeing the massive investment being made (estimated by industry observers at c. $200 billion), innovation in technology, education and awareness building with consumers, government intervention and support (especially in the USA by President Clinton and Al Gore) and the necessary experience and expertise gradually being established in operating and marketing electronically.

As we shall see, it's no longer a question of whether electronic shopping will come to change our lives. It's now simply a matter of when, and to what degree. When will it reach critical mass? Is it within existing planning horizons or can it be ignored and shelved for a few more years yet? Who will be most affected? Are all retailers vulnerable to electronic shopping trends? Will they simply be disintermediated or can they respond and not just survive but build on this new opportunity? What about manufacturers – how can they exploit the situation? And consumers – to what extent are they interested in shopping in this new way, will they *all* eventually become dedicated electronic shoppers?

There's so much change going on, the impact is so wide-ranging that we need an early summary of the specific messages and themes in this book to set the scene and establish the context. This chapter sets up the main ideas and summarises what amounts to the thirteen major research findings and insights. They are introduced here by way of initial sound-bite and brief illustration only. They are then explained in detail with case studies and the research more fully examined through the subsequent chapters of the book.

These are the key findings and conclusions:

1. The revolution in electronic shopping *is* going to happen
2. It's going to have a major impact on the shopping and retailing scene
3. Already c. 15–20% of consumers say they'd prefer to shop electronically rather than visiting the shops
4. It only takes a drop of c. 15% in store traffic to make many stores unprofitable
5. This revolution will achieve critical mass by as early as 2005
6. As store visits reduce, some shops are in danger of dying out; mid-sized high streets and malls are the most vulnerable
7. From supermarkets to banks, nearly all retailers will be affected
8. Companies should quickly assess the potential impact on their business using the 'ES Test' set out in Chapter 5
9. Not all is lost! Retailers can survive and succeed but need to decide which of the ten identified strategic options they are going to pursue
10. Manufacturers can seize the opportunity to decide whether to establish their own direct consumer distribution and bypass existing retail chains
11. The electronic environment will demand new marketing skills
12. Future success urgently requires the development of a clear long-term strategy . . .
13. . . . and rigorous implementation and communication.

Let's now consider in a little more detail each of these key messages (research sources are listed in the bibliography):

## 1. The revolution *is* going to happen

All the evidence points to an unstoppable momentum, an inexorable force that will drive electronic commerce forward and reach out to make it so pervasive and accessible that it can't fail to impact shopping habits:

## Overview: Key Conclusions

- 'President Clinton has announced his personal support for a five year programme to develop the Internet as a universal medium open to everyone. The White House will fund the program in its first year with a near $1bn commitment'
  *(Reuters News Service, 1997)*

- VP Al Gore is targeting every classroom in the country to be connected to the Internet by the year 2000. 'It's an enormous effort, comparable to building the nation's roads and railways but it's essential.'
  *(Reuters News Service, 1996)*

- 'A new generation is growing up digital. By the year 2000 there will be more than 88 million people in US and Canada between the ages of 2 and 22. They are the kids who are leading the charge using the new media that's centered around the Internet. This Net Generation is developing and imposing its culture on us thereby reshaping how society and individuals interact.'
  *(Don Tapscott, author of 'Growing up Digital')*

- 'In 1998, 22.8 million US households will be wired to the Net.'
  *(Business Week)*

- '30% of households in the major Asian economies will be connected to the Internet by 2002.'
  *(NUA Internet survey, 1996)*

- 'Communication bandwidths will rise exponentially and hardware equipment will become simultaneously more powerful, sophisticated, easier to use, affordable and portable. Once the hardware and telecommunications infrastructure has been established, an enormous range of services can be exchanged at nominal incremental cost, such as location-independent shopping and banking.'
  *(Jagdish Sheth, Professor of Marketing, Goizueta Business School, Emory University)*

- 'There's no doubt that over the next five years, electronic shopping will become an important facility for customers. Too many people are investing too much money for that not to happen.'
  *(VP of former Burton Group, UK $4bn clothing retailer)*

## 2. It's going to have a major impact on the shopping and retailing scene

Commentators on the marketplace have no doubt about the potential. It's not going to be just another distribution channel or an experience confined to 35-year-old male computer nerds. It's going to become a mass medium for communication, interaction and all forms of buying and selling:

- The Internet is a tidal wave. It will wash over nearly all industries drowning those who don't learn to swim in its waves.'
  
  (*Bill Gates, CEO of Microsoft*)

- 'It won't be too long before 20 to 30% of households will be buying electronically. This will mean boundless potential for bypassing traditional distributors.'
  
  (*Andrew Grove, CEO of Intel*)

- 'Retailing will never be the same – home shopping will revolutionise retailing'
  
  (*Business Week*)

- 'Virtual stores are the future. They are the natural extension of the ECR [Efficient Consumer Response] initiative to streamline the logistics of getting products from manufacturers to consumers.'
  
  (*Raymond Burke, Professor of Business Administration, Indiana University*)

- 'One third of consumers interviewed said they expected to do most of their shopping in the future via the telephone, Internet or television.'
  
  (*Bossard Consultants research, 1997*)

## 3. Already c. 15–20% of consumers say they'd prefer to shop electronically rather than visiting the shops

Consumer research shows significant numbers of people instinctively appreciate the benefits of shopping electronically. Many have busy lives and struggle to find enough time to do things. Shopping is often a hassle and regular purchases are commonly seen as a chore. In this era of time poverty many would prefer to have the opportunity for leisure, sport or relaxation. The underlying convenience of electronic shopping is just too tempting to ignore:

- '– 13% of consumers actively hate going shopping
  – In total 16% of consumers said they 'generally dislike shopping'
  – As many as 31% of all consumers would classify themselves as "reluctant shoppers" finding the experience "not really enjoyable".'
  
  (*Mintel research, 1996*)

- '19.4% of all shoppers said that "if there was a grocery shopping and delivery service in my area I would use it".'
  
  (*A. C. Nielsen research, 1997*)

- '32% of people who say they don't have enough time to go shopping – a group classified in the research as "frenzied copers" – said they would be interested in home delivery.'
  
  (*A. C. Nielsen research, 1997*)

- 'Consumer acceptance of non-store retailing is already high. 45% of US households and 58% of UK households purchase products from catalogs or buy via television each year.'
  
  (*Verdict Home Shopping report*)

## 4. It only takes a drop of c. 15% in store traffic to make many stores unprofitable

Retail margins are often wafer thin. 2–3% is quite common and UK averages of 5–6% plus are not the international norm. In these circumstances retailers are very vulnerable to relatively small shifts and reductions in the number of people who visit and shop at the store:

- 'It will only take a 10–20 percentage point shift in the retail market from the high street to home to eliminate most retailers' profit margin.'
  
  (*Goldman Sachs research*)

- 'Even a 10% loss in consumers visiting our stores would be a worry. For many of our outlets that would push us below breakeven.'
  
  (*Managing Director of $2bn plus retail group*)

- 'Average retailer margins will be eliminated by a 15–20% reduction in consumer traffic through their stores. It is far from clear that this margin loss can be made up by capturing the electronic purchasing instead. Consumers may buy from an alternative provider.'

    *(CSC Kalchas research)*

- 'Even a small shift in purchases from traditional to virtual stores will have dramatic implications for retailers and manufacturers.'

    *(Professor Raymond Burke)*

## 5. This revolution will achieve critical mass by as early as 2005

Forecasters vary widely in their predictions on the growth of electronic shopping. Many foresaw an explosion of interest reaching significant levels by as early as the end of this century. Since the initial hype more measured assessment has taken place. Most now agree that it is in the next decade that the real growth will come. It will happen as the electronic environment begins to move out of its experimental phase into its first era of maturity. The infrastructure will be much more widespread and technology will have reached a greater level of sophistication offering much easier to use yet higher quality electronic communication Companies will have begun to learn how to make real the convenience and added value of electronic shopping:

- 'By the year 2000 a third of the most attractive households will be ready to buy on line.'

    *(McKinsey)*

- 'The combination of technology sophistication, equipment power and ease of use plus the supporting infrastructure will make electronic purchasing widespread in the US by the year 2005.'

    *(Jagdish Sheth, Professor of Marketing, Goizueta Business School)*

- 'According to the CBI [Britain's major industry association], the Internet technology available in 10 years [2007] will be sufficient and widespread enough to outstrip the high street in terms of sales.'

    *(Internet Business report, 1997)*

- 'There will be a new wave of growth around the middle of the next decade responding to new user-friendly easy access low cost

## Overview: Key Conclusions

technology. It will enable the 15–20% hard core of "reluctant shoppers" who dislike going to the store to become near dedicated electronic shoppers instead.'

*(CSC Kalchas research)*

- 'Electronic retailing will grow inexorably over the next decade. By 2010 it may account for as many as 25% of US retail sales.'

  *(Professors Louis W. Stern and Barton A. Weitz)*

- 'Non store retailing could account for as much as 55% of "total retail sales" by 2010.'

  *(Kurt Salmon Associates)*

- 'In 30 to 40 years time (around 2030) there may be no shops at all.'

  *(Maurice Saatchi)*

### 6. As store visits reduce, some shops are in danger of dying out; mid-sized high streets and malls are the most vulnerable

As it gets easier to purchase electronically so by definition the number of store visits will reduce. When consumers do go out to shop the basics will already have been delivered. They'll be looking for a new type of shopping experience. Expectations will be different. They'll demand new things, a wide range of choices, the opportunity to relax and enjoy themselves combining that with other leisure, social and entertainment activities. In that regard the big supermarkets, the larger malls and high streets have more space and more flexibility to respond. Small high streets and shopping areas just won't be able to meet consumers' widening needs and expectations:

- 'A quarter of British people believe the traditional local high street is bound for extinction because of the growth of home shopping.'

  *(Bossard Consultants research, 1997)*

- 'The Internet will drive half of today's retailers out of business.'

  *(Research by International Business Development Corp.)*

- 'By 2010 there may not be much left of the high street.'

  *(Eat Soup magazine)*

- 'Shifts from traditional to virtual stores will have dramatic implications. I expect you'll have left warehouses and then you'll have 7-11s and that would be it. All the major stuff would come out of these central locations. You'd be interested in the cheapest cost for the product that you want. Then you go to the convenience store for the last minute things that you need.'

*(Research input to Professor Raymond Burke)*

- 'Consumers are clearly looking for a way to divide shopping for staples distinct from browsing for fresh/novelty/impulse/luxury purchases.'

*(Financial Times Management report)*

## 7. From supermarkets to banks, nearly all retailers will be affected

The electronic shopping phenomenon is not restricted just to grocery. It cuts across all retail consumer sectors with financial services one of the most vulnerable areas. Basic banking and insurance have already proved popular sellers not just on the Internet but simply by using phone or fax. The number of new entrants dedicated to electronic selling is growing rapidly and can be found in just about every area of consumer activity:

- 'Products and services bought on-line range from software as currently the most popular through to books, music, electrical and electronic goods, food and clothing.'

*(Verdict Electronic Shopping report)*

- 'Clothing sales already dominate mail order shopping, evidence that clothing can also succeed on line and that consumers are prepared to buy *before* trying.'

*(Verdict)*

- 'Consumers are flocking to us and some of our most popular food lines are fresh fruit and vegetables ahead of other packaged goods.'

*(Peapod, a US grocery/food distributor offering electronic shopping services)*

- 'Direct Line (the UK insurer) have demonstrated the attractiveness of selling insurance over the phone. Their fast rise to market leadership shows consumers are happy buying over the wires.'

*(Internet Business 1997)*

*Overview: Key Conclusions* 15

- 'First Direct telephone banking is growing new customers at more than 20% per annum and many are now describing their banking as a "pleasurable experience".'

  (*Marketing Week, 1997*)

- '$8bn worth of entertainment and travel tickets will be bought on line in the US in 2001.'

  (*Jupiter Communications research*)

- 'Computer hardware company Dell reports selling $3 million-plus worth of computer products a day over the Internet.'

  (*Forrester Research, 1998*)


### 8. Companies should quickly assess the potential impact on their business using the 'ES Test' set out in Chapter 5

In Chapter 5, the 'ES (Electronic Shopping) Test' is described. It sets out a framework to decide which products and services are most susceptible to being sold electronically. The test provides a guide to retailers and suppliers. Are they in a product/service field which will be untouched by the electronic revolution? Or are they operating in an area which will be particularly attractive to the would-be electronic consumer? Is there a need to change urgently or can they simply wait and see? The ES test offers a simple and practical way of determining the impact of these new market forces. It has three components examining: (1) Product characteristics of touch, taste, smell, etc; (2) Consumer familiarity and confidence with the particular product/service and (3) individual consumer interest and demand for 'convenient shopping':

- 'Products which traditionally need to be **touched, tasted** or **smelled** are prima facie less likely to sell well on line. But products and services without those characteristics or where they are of low importance will have electronic appeal.'

  (*CSC Kalchas research*)

- 'Where consumers are **familiar** with the product or trust the brand name they'll be more comfortable buying by phone, PC or TV.'

  (*Professor Sunil Gupta, University of Michigan*)

- 'The search for **convenience** will be a major driver. The modern consumer is often described as overworked, stressed and time-poor. But they are demanding. Those for whom convenience is important will respond to the virtual store that can deliver satisfactorily.'

*(FT Management report)*

### 9. Not all is lost! Retailers can survive and succeed but need to decide which of the ten identified strategic options they are going to pursue

Research and testing has shown there are up to ten alternative strategies that are available to retailers in choosing their response to the electronic revolution. In Chapter 7 these ten strategies are described and evaluated. They range from a relatively low-key response through to aggressive investment in the new medium. But there are also options to 'buck the electronic trend' and reinvent the role and value-added of the physical retail estate:

- 'Retailers are allegedly frightened by the implications of responding to the threat of electronic shopping. If they set up their own web sites will they simply cannibalise their store sales? . . . they have made massive investments in their physical infrastructure. Should they now jeopardise all that and undermine long-established customer relationships?'

*(Verdict)*

- 'Retailers can continue to develop and transform their existing retail channels. For example the four main UK grocery multiples have aggressive store expansion plans. These will provide competition to the Internet.'

*(FT Management report)*

- 'Retailers must decide what are the innate characteristics for their products or services and whether they are easily susceptible to being bought electronically. With that input they can decide whether they need ultimately to abandon their real estate or whether they can still make it work.'

*(CSC Kalchas research)*

*Overview: Key Conclusions*

- 'Retailers do have the choice whether to be active or passive. Either choice has however to be creatively and comprehensively followed through.'

  (*Professor Fred Phillips, Oregon Graduate Institute*)

## 10. Manufacturers can seize the opportunity to decide whether to establish their own direct consumer distribution and bypass existing retail chains

Manufacturers have spent the past twenty years dominated by their retail customers. They have been pushed for bigger discounts, squeezed on shelf space, seen private label initiatives copying their own product innovations even down to similar looking packaging. Electronic shopping presents them with an opportunity to fight back and re-establish their links with their end user. Rather than continue to struggle with retail distributors they could deal direct. In evaluating their options a further set of ten alternative strategies, this time tailored for manufacturer suppliers can be identified. They range from 'sticking to their knitting' through to establishing their own home delivery operation as their principal means of distribution:

- 'If your company deals through agents, wholesalers or retailers it's time to do some serious strategising . . . All those roles can be disintermediated . . . food producers won't need supermarkets when customers can replenish supplies weekly by accumulating entries in their shopping list database and take delivery at home . . . hotels won't need travel agents.

  Take the case of consumer goods manufacturers being squeezed by giant retailers like Wal-Mart. They could use the new infrastructure to sell direct over the Net. An electric tool and appliance company like Black & Decker could provide interactive programs for a fee on say home renovations featuring their tools and offer an immediate purchasing facility.'

  (*Don Tapscott, The Digital Economy*)

- 'Manufacturers are reacting to the pressure tactics of traditional retailers by seeking direct links to end-users through the Internet and mail order operations . . . Apparel and accessory designers such as Donna Karan, Giorgio Armani and Liz Claiborne . . . Nike and Sony . . . Airlines for example, are letting consumers book reservations bypassing travel agents . . . many consumer products manufacturers have created sites on the Web, hoping this link to consumers will ultimately permit them to reduce or eliminate their dependence on retailers and dealers.'

(Professor Nirmalya Kumar, IMD)

- 'Some manufacturers will start building up their direct capabilities (Procter & Gamble and Unilever are two examples) and retailers will have to play to their strengths if they are to compete'

(FT Management report)

## 11. The electronic environment will demand new marketing skills

Consumers will have much greater choice. They won't be dependent on buying from those shops they can physically access within reasonable drive times. They can buy whenever and wherever with a choice all over the globe. In addition, electronic shopping offers them the chance to set up a very personalised relationship with their Internet or other electronic provider who can respond by tailoring their offerings to individual segments of one. Marketing will have to become more selective and segmented, general mass advertising and communication may be too generic and miss the target. But brand awareness will remain vital. Guiding consumers to the specific web site will be a new challenge and making it easy once they get there the most critical point of leverage:

- 'Marketers who are unprepared for this sea change we're about to experience won't survive. But those who do grasp the implications and get ahead of the curve will emerge more competitive. They'll be forced to build relationships with consumers that are deeper and more enduring than any we can create today.'

(Bob Wehling, Senior VP at Procter & Gamble)

- 'Companies still have a lot to learn about effectively marketing on the Web. Currently it is hard to find retailers' web sites. They can only work if customers know exactly where to go to find the products they want. Try looking for "retailers of groceries" in the web's best search engines and you're going to be disappointed ("no matches" is the response). Try searching for "food and detergent in Birmingham" and you will suffer the same fate. Being in exactly the place where most consumers will look for what you sell is the essence of location,

Overview: Key Conclusions 19

location, location. Being available on line but impossible to find falls a long way short of that.'

(*Verdict electronic shopping report*)

- 'The expanding global electronic marketplace has powerful implications. The introduction of interactive media and the on line environment has made real-time, customised one-to-one advertising, marketing and commerce possible. Interactive multimedia can no longer be considered an afterthought when developing an integrated marketing campaign.'

(*Margo Komenar, Electronic Marketing*)

- 'Almost 30 years ago, AT&T pioneered the first steps in electronic marketing with the invention of the 800 number. Now the merger of the communications and computing industries is transforming the way the world works, sells and gets its information and has created a platform on which marketers can create new value for their businesses.'

(*John Petrillo, Executive VP, AT&T*)

- 'The web gives us an immediate global reach but also the opportunity to customise what we offer to a market segment of one.'

(*Jeremy Silver, Virgin Records*)


### 12. Future success urgently requires the development of a clear long-term strategy...

This new market challenge in one respect is no different from any other. Success can be achieved only if companies confront it and determine to work on developing a clearly thought-through plan of action. Immersion in the market place, understanding customers' needs and wants in a comprehensive and rigorous way, identifying the various options and evaluating which will have the best leverage – these are all critical steps any current player in the retail scene must now go through:

- 'Companies that plan their way through emerge as winners. "Visionary companies" outperformed their industry rivals on average by a factor of six on the basis of stock market returns on investment.'

(*Collins and Porras*)

- 'Nowadays we are having to make ever bigger bets in our investments ... we can't afford to back one direction and then find out five years later it was wrong ... so strategy is our most important management issue.'

  (*Chairman of United Parcel Services*)

- 'The market place has become saturated with competition. New products and new market channel opportunities are emerging with increasing frequency in every industry. Mapping the way through this market maze is an enormous challenge. Winning companies are ones who are seizing that challenge firmly in both hands and with every effort. They are developing a long term plan and vision of what they want to achieve and in detail how they are going to get there.'

  (*Foreword to Strategy in Crisis, 1997*)

- 'Key to growth is the development of a clear, long term strategy.'

  (*Coopers & Lybrand/Sunday Times research*)

- 'Those that do well in the electronic market place will be those who best understand their customers ... and develop an effective game plan to evolve and meet their needs.'

  (*Professor Joseph Hair, Institute of Business Studies, Louisiana State University*)

### 13. ... and rigorous implementation and communication

Strategic thinking and planning, setting the framework and plan for the future – that's the first half of the equation. Next comes the most difficult task of pursuing it forcibly, getting the workforce to buy-in to the targets, feel motivated and involved and wanting to go the extra mile, then being able to stick to targets and hit milestones – all part of the challenging path of making things happen and adapting to change. For some companies this may result in significant upheaval and new investment in systems and skills. But given the way electronic commerce is moving and the speed of its development there may be no choice:

- 'Corporations are learning that handsome returns come directly from a superior combination of strategic planning and strategic execution.'

  (*Professors Bartlett and Ghoshal*)

## Overview: Key Conclusions 21

- 'Effective business leaders of today communicate a compelling and believable strategy and vision of where they want the company to go and understand what role and responsibility their workforce has in getting the company there.'

  *(Charles Handy)*

- 'If people are to put out the extraordinary effort required to realise corporate targets then they must be able to identify with them and share in the ideas and goals they represent.'

  *(Jack Welch, GE)*

- 'All the things in the value chain need to work effectively to deliver the best possible experience for our customers ... to add customer value a firm cannot just sell a product, it must also service what it sells.'

  *(Kenneth Hill, VP, Dell Computer Corporation)*

- 'Existing mail order companies have a head start in electronic commerce. They already have a well-developed highly experienced infrastructure experienced in dealing with customers remotely, handling any complaints, packing goods safely and delivering reliably. In effect they have developed a high performance fulfilment operation which will set a benchmark for others moving down this path.

  *(FT Management report)*

\* \* \*

Electronic commerce *is* clearly going to be part of everyday business life. It will be one of the primary dynamics as the first decade of the twenty-first century unfolds. Even today the pace of change is extraordinary. Every week there is a new improvement being announced in the technology, equipment is getting cheaper and the reach of the Internet is widening through TV as well as PC access. Consumer life-styles are becoming if anything more pressurised for time not less, thus accelerating the search for convenience and the need to find enjoyment in what little spare time can be carved out.

The present relatively low levels of purchasing on-line cannot be used as an excuse for inaction. They are a temporary state of affairs while the technology improves and the investments poured in during the mid-to late 1990s begin to bear fruit. The analysis described in this book shows that the middle of the next decade will be a key period in the development of electronic buying and selling. 2005 is the date being pinpointed. By that stage many of the experimentation and teething problems will have been sorted out. Business generally will be in a far more capable state to exploit the new medium and find new sources of advantage and business growth with target customers.

No business can afford to stand still. The 'ES test' can be used to pinpoint the extent of the 'danger' – but also the opportunities. The 'ten strategies' evaluation can be reviewed to help fix on the most appropriate and leveraged course of action. As always the winners will inevitably be those who grasp the nettle most firmly, develop the most rigorous understanding of their customers and end-consumers and be so in touch with their market places that they can almost anticipate what initiatives and moves will work. Combine clear planning with rigorous implementation – in theory at least a straightforward formula – and the potential is there for growth and success in the electronic marketplace.

\* \* \*

**Structure of the book**

Following this Introduction and overview, Chapter 2 stands back for a moment to set some context for this electronic shopping review. Some definitions need to be made for the terminology employed and some emphasis and priority described in considering different industry sectors.

Chapter 3 is one of example and illustration. There's a lot going on in the electronic arena. Traditional mail order houses are not standing still and are rapidly developing their on-line capabilities. Many manufacturer suppliers are quietly but determinedly establishing their web sites and learning how to market over the Net. The retail community is at the forefront of many new initiatives and are trialling electronic ordering. New entrants, unencumbered by existing physical infrastructures, are emerging with exciting and dedicated on-line offerings. What's the extent of all this activity and what learnings are already accumulating in the market place?

Chapter 4 examines the evidence and looks to pinpoint when electronic shopping is likely to reach critical mass. While the USA is some way ahead, the UK and Germany are setting the pace in Europe and the major Asian economies are not far behind. The middle of the next decade, c.2005, is likely to represent an important stage in the evolution of the technology and the infrastructure.

How can a retailer or manufacturer analyse if its products and services are vulnerable to these forces of electronic shopping, and if so to what degree and how immediate is the threat? Chapter 5 sets out the 'ES Test'

## Overview: Key Conclusions

which provides a framework for defining the need and extent of the response required.

In this new era, the challenge facing established retailers is enormous. But they have had a history of successful evolution. They have been pioneers of developments in the supply chain. Through self-service in the 1950s through to hand-held scanning and bar code terminals they have themselves fundamentally changed consumer shopping habits and expectations. Are they now equipped to respond adequately to electronic shopping developments? Does their leading position in the distribution chain make it easier for them to evolve and retain the consumer relationship? What kind of platform do they have to build on in responding to these new challenges? This is examined in Chapter 6.

The next Chapter, 7, sets out the range of responses retailers can choose from. There are ten alternative strategies that can be defined. The message is that no retailer can afford to ignore these market developments.

One strategy moving forward is for retailers to stick with their real estate but transform it so consumers still want to visit, despite the home shopping temptations. The successful 'store of the future' – Chapter 8 – will need to represent a completely new experience. It is unlikely to be just about getting the goods off a shelf and paying for them at the check-out. If the real estate is to continue to add value, retailers must rethink the way the space they have can be utilised. How can it meet changing needs and expectations? How can it be evolved to provide a new and exciting experience? Can major supermarkets, for example, establish and capture some role that continues to lock them firmly into the hearts of their 'local community' and catchment area? In fact, can they achieve an even more entrenched position? We can illustrate by looking at one kind of 'store of the future.'

Chapters 9–11 set out in more detail what's driving the developments of electronic shopping and why it appears unstoppable. Executional difficulties such as individual privacy and security of payment are gradually being resolved. Indeed the view has been expressed that 'payment for goods over the Internet will soon be more secure than the regular use of credit cards for ordering over the phone or in the mail'. Chapter 9 especially looks at how the technology and infrastructure are moving forward and how improvements in what can be transmitted over ordinary telephone cable, the possibility of using the electricity grid and developments in satellite communication will all accelerate reach and penetration into more households as well as enhancing the visual quality of what can be communicated.

In Chapters 12 and 13 there is a review of how manufacturer suppliers could now look to take advantage of the new medium. Like their retail counterparts they have a range of options and a further set of ten alternative strategies can be defined and described.

The final Chapters 14 and 15 consider what it will take to be a winner in the new electronic environment. Responding to the changing scene will require investment and carry risk. Some tough decisions will inevitably have to be made, some sacred cows abandoned. Key to determining the most leveraged response will come from being market and customer rooted, so understanding the market place that the company is totally in touch with its customers. Rigorous assessment will help determine whether and how critical it is to respond electronically. Market opportunities can be prioritised and clear plans and strategies pinned down. As understanding crystallises so companies can better determine when to start changing and how best to transition from where they are to where they will need to be.

Mapping the path through the new market challenges is going to be critical. Clarity in thinking and in direction-setting will naturally help the workforce appreciate the new direction, get them more involved and motivated behind the new targets and more committed to go the extra mile to achieve them. Electronic commerce is likely to be one of the most fundamental changes business will face and there will be an additional premium on putting the necessary time and resources into developing the best response and way forward.

# 2 Definitions and Consumer Trends

Some brief definitions and explanations are required before moving into the detail of what retailers and manufacturers are doing about the electronic revolution and how they can best respond.

This chapter looks at:

- a definition of electronic shopping
- the products and services included
- underlying themes behind all the changes.

## Definitions

In defining the various terms and expressions, it's important to immediately set out what is meant and included in 'electronic shopping'. In fact it incorporates a wide range of different media or what can be described as different electronic connections (Table 2.1). The Internet is but one of these, though emerging as by far and away the most prominent. (For reference the term 'electronic shopping' will be used frequently through the book and will sometimes be abbreviated to 'ES'.)

**Table 2.1** Electronic Shopping Connections

| | |
|---|---|
| • Communication/access devices | TV, radio, PC, phone, fax, ATM, interactive kiosk, pager, personal digital organiser |
| • Content media | catalog, CD Rom, teletext, world wide web, terrestrial and satellite broadcasts |

## Products and services

These 'electronic connections' cover a wide range of commercial transactions but the focus here is on the consumer. The aim is to look at the interface and interaction between retailers and end-users, how that might be changed by these new market forces and the role and opportunity for manufacturer suppliers in that context.

Consumer shopping covers almost every retail product area and a number of specified retail services. The detail is set out in Table 2.2. In principle, all the areas listed will be exposed to some electronic selling and buying and so need to be investigated.

**Table 2.2** Consumer Products and Services

| | | |
|---|---|---|
| **Products** | • Grocery including meat, fruit and vegetables<br>• Clothing<br>• Footwear<br>• Music<br>• Books<br>• Computer hardware, software<br>• Household appliances<br>• Brown and white goods (e.g. HiFi and washing machines)<br>• Sporting goods<br>• Camera and photographic supplies | • Confectionery<br>• Bakery<br>• Home improvement and building materials (DIY)<br>• Garden products<br>• Furniture<br>• Furnishings<br>• Motor vehicle and bike retailing<br>• Auto supplies<br>• Pharmacy and OTC (over the counter) |
| | – Eating and drinking places are excluded | |
| **Services** | • Retail banking<br>• Personal insurance<br>• Financial advice<br>• Stocks/mutual funds<br>• Travel<br>• Holidays<br>• Entertainment tickets<br>• Home education | • Real estate<br>• Consumer information<br>• General information services |

The grocery sector will often be used to provide examples and illustration. This is for a number of reasons:

- Grocery sales are a substantial part of all retail sales. In the USA some $400bn is spent on grocery, about one-fifth of total sales. In the UK the proportion accounted for by grocery is closer to 25%.
- It is estimated that around half of the world's population is involved in the production, distribution or sale of food products.
- Time spent by consumers in buying grocery products is considerable, on average making 2.5 trips to the store each week.
- Research shows c.80% of items bought are simply replenishment purchases (e.g. milk, bread, paper towels, soft drinks).
- According to a 1990 University of Michigan study, consumers ranked grocery shopping as one of the least enjoyable daily activities!

## Underlying themes

A number of case studies will be discussed through the book. They will illustrate and reflect a set of underlying trends in the market place which are fuelling not only the development of electronic shopping but also a transformation in commerce and daily life. These underlying trends (Table 2.3) will often be referred to in more detail but some brief introduction and explanation is also required early on.

**Table 2.3** Key Underlying Trends Behind Electronic Shopping

- Consumer time poverty
- Consumers looking to take control
- Convergence of technologies
- Shift from physical to digital
- Shift from assets to knowledge

## Consumer time poverty

In today's pressured environment it has been well documented that people suffer from 'time poverty'. There is luxury of choice but chronically less time to indulge. 'Spare time' is more precious and consumers are constantly searching for products or services that help save time, improve life-styles or make it easier to do things. Shopping electronically once it's made easy may be just what consumers having been waiting for.

## Consumers looking to take control

In the context of time poverty, consumers don't want to be dependent on shop opening hours or physical location of the retailers. They want to be able to buy what they want when they want it. As *USA Today* pointed out in a recent article, we are moving into an 'I want it now' society. Consumers of today are now better educated and more sophisticated than their predecessors. They are more demanding, expectations of service and convenience are high and they search restlessly for new ways of meeting these needs.

**Convergence of technologies**

The overall structure of the economy is changing. A new industrial sector is emerging from the convergence among computing, communications and content companies. It's creating a multimedia industry of more than $1 trillion with computing companies leading the way. It's the combination of these technologies and skills, the alliances and partnerships between companies from the different sectors and the billions of dollars of investment that is going to revolutionise what can be achieved in electronic commerce.

**Shift from physical to digital**

Information flows used to be all physical, typically on paper such as cash, cheques, invoices, catalogs and and shopping lists. In hard copy form transmission and processing times are slowed. Now the information superhighway is seeing the conversion of what was physical into bits stored in computers that are inter-networked across the globe and can communicate instantly.

**Shift from assets to knowledge**

Retailers' current business is built upon their assets. Their leasehold or freehold sites are often in high-cost real estate and interiors have been expensively kitted out. But the new electronic economy is based upon knowledge not assets. It's about the exploitation of information and know-how without reference to physical location. Now companies can enter the market place and transact with consumers without the expensive infrastructure. It's going to put increasing pressure on existing real estate distributors.

* * *

These terms and definitions form part of the basic lexicon of electronic shopping. There's a whole new world taking shape and as it does so a new vocabulary for commerce will develop with it. We will all become increasingly familiar with dot coms, web sites, interactive media, search engines and browsers. We will all gradually become wired up and on-line and in touch with a new electronic age. It's already a recognisable and fashionable *lingua franca* among the new generation of Net kids and others are latching on. The general mass media of press and TV is typically quick to embrace and popularise new ideas and soon large numbers of consumers will be going 'electronic'.

# 3 Learning from the Pioneers in Electronic Selling

The retail shopping scene is already changing. At the leading edge it's moving from casual flirtation with electronic shopping to much more serious investigation. It's developing in profile from a pet project for the marketing department to an opportunity that the whole organisation (especially in the areas of marketing, logistics and systems) will need to come to terms with and integrate into future plans.

There are a number of companies who are beginning to emerge as leaders in the new arena, pioneering change in their industry sectors and looking to establish new forms of advantage and differentiation in their market place. They see electronic commerce as exciting, not as a threat. They are innovating and ready to take 'constructive risks'. They are proactively investing and investigating. They want to be recognised by their customers as being at the forefront of their industry and doing all they can to strengthen their product service value proposition.

Who are these industry pioneers and adventurers? How are they changing the rules in their sectors? What are their strategies, what lessons can be learnt for others who are looking to come in and explore these new electronic opportunities? This chapter investigates those companies who are among the most innovative and proactive. It also considers the impact already being achieved through the general level of trial and experimentation.

**Industry pioneers**

There are some eight corporations who stand out in the field of consumer electronic commerce. Several of these did not need the advent of the Internet to awaken their interest. They were already tapping into consumer needs for convenience and added value by exploiting very successfully the 'old-fashioned' electric connections of phone, fax or TV. Others have taken the Internet as their catalyst and are at the forefront in their direct to consumer interaction.

The eight companies which deserve the accolade of 'pioneers' are:

- Direct Line
- 1-800-FLOWERS
- First Direct
- QVC
- Dell
- Levi's
- Amazon.com
- Tesco.

These eight companies will now be discussed in turn.

*Direct Line*

Direct Line was established in the UK in the mid-1980s and in the ten years of its existence has rocked the insurance industry, driving former market leaders into share decline and profit loss. It has changed the way consumers buy auto and household insurance. Insurers in other countries have watched these developments with growing anxiety and many have been investigating what strategy they might adopt if and when the same phenomenon takes place in their own domestic arena. Direct Line's former CEO Peter Wood is planning an assault starting in 1998 on the US market, focusing in on certain states and consumer segments and threatening to revolutionise the US domestic insurance market in the same way as has happened in the UK.

What did Direct Line do that was so revolutionary? Simply, it dealt direct with the consumer. The company advertised on TV and in the press and encouraged consumers to phone or fax in. By cutting out the middle agents, brokers and distributors it saved costs on infrastructure and third-party profit share. It translated those cost savings into aggressive advertising to build awareness and supplemented that with sophisticated consumer targeting techniques, eschewing high-risk policies and majoring on age groups with safer track records.

By the mid-1990s, just around seven years after launch Direct Line had achieved UK market leadership. Its competitors, initially slow to respond on the basis this would be 'just another marketing channel', have all now rushed to set up their own direct operations. They have realised that it's no longer just about alternative channels – 'direct' has now become the principal way of selling and buying personal lines insurance.

What were the reasons behind Direct Line's success? Its key was to fully fund and resource its market initiative. It realised it had to change consumer buying patterns, it recognised that that was a tough but feasible target in the market circumstances and that it would need to create high levels of awareness and reassurance. As a result there was considerable investment in advertising building up the brand and establishing an accessible and friendly direct phone operation with consumers. Direct Line knew that half-hearted efforts in this direction would lead to failure. It had to break through the noise barrier. It had to break the mould if it was to succeed. What was also in Direct Line's favour was that it could enter this market greenfield. It had no established branch and agency infrastructure resisting the change and pulling it back to 'old ways' of doing business. It could set its sights, exploit its full investment and inexorably grow its direct operation.

Direct Line's rivals, the long-established insurance groups, have not had this greenfield luxury. While they have not sat idly by, many have struggled to make their direct initiatives a success. They find themselves stuck with conflicts of interest – established intermediary channels pulling one way, 'direct to consumer' opportunities pointing them on an alternative path. A main challenge now is learning to manage this 'channel conflict': how to take advantage of the new market opportunities without totally cannibalising or undermining existing operations.

Can that be achieved? There is learning and experience to suggest it can and this will be examined in detail in Chapters 6 and 12. But the key will lie around segmentation, around so understanding different customer needs and desired ways of purchasing that a targeted differentiated marketing and distribution programme can be confidently and persuasively pursued.

## 1-800-FLOWERS

1-800-FLOWERS is another electronic success that did not need the Internet to get started. It has built a flourishing business on the back of the 1-800 free phone number ('the first truly electronic marketing tool' according to its inventors at AT&T).

Its goals have been to build a fast but friendly service operation 'making human contact and connections'. It has grown rapidly ('we didn't need to build a nationwide chain of stores'). Since it was set up in

1985, 1-800-FLOWERS has become the world's largest florist. It has more than $300 million in sales and over 2000 employees.

Jim McCann, the company's President, has often been described as a pioneer ready to 'break business rules'. He realised that a major challenge would be to get people comfortable with buying flowers over the phone, when they couldn't see, touch or smell the product beforehand. With that in mind he established and heavily promoted the '100% Quality and Assurance guarantee' that promised the flowers would be fresh and last a full week or money would be refunded. The company went out of its way in its early days to build up consumer comfort and confidence. It added follow-up telephone surveys to check on satisfaction with both customers and recipients and set up a substantial programme of local education classes about flowers as a way of finding out what customers wanted, what was working and what needed improvement.

Once the Internet started developing as a commercial network, 1-800-FLOWERS was in a prime position to take advantage. It had the electronic shopping know-how. It had the infrastructure in place. There has been little room for any new comers to try to usurp the company's position as the 'electronic seller of flowers'. In its first year of marketing its web site, 1-800-FLOWERS added an additional $30 million to sales. 'We are truly interactive . . . our stores on the web are as important to us as brick and mortar retail stores in local markets. But we have found that people consider us a more convenient location.'

*First Direct*

First Direct Bank (a UK subsidiary of the Hong Kong and Shanghai Banking Corporation) has been a patient and steady developer of telephone banking services, following in the footsteps of leader Citibank in the USA.

While rivals stood by or offered basic call-centre services, First Direct has built up a sophisticated telephone and PC banking operation and a customer base of nearly 1 million UK households. It offers a low cost 'bank when you please service'. it's particularly targeting the 'frenzied coper' (see Chapter 9), the busy executive or 'person on the move' who doesn't have time to visit the branch between 9.30am and 4.00pm Monday to Friday.

Its success lies not just in the convenience of its service but most particularly in the way that service is carried out: 'it makes banking a pleasure', 'I can call up any time and they always answer immediately and are friendly and efficient', 'they've got a good database and know what I need'. It is principally a result of their service excellence that their reputation and business is growing. 'We're currently adding new customers at the rate of 12,000 a month and much of it is word of mouth recommendation.'

Other banks competing in the UK market have examined setting up rival operations and some like Lloyds TSB and Barclays have taken steps to do so but their initiatives have often been too tentative. They, like the insurance companies, are struggling with the conflict around channels of distribution. Would encouraging their customers to bank by phone, fax and mail or via their PC simply accelerate the demise of the physical retail branch? If that happens does the bank lose its value-added role in the community? Can that be sustained or replaced in a 'virtual' selling environment?

These questions are growing in importance as electronic commerce develops and their resolution cannot be postponed indefinitely. The closer we get to the 2005 'critical mass date' the more important it will be to build up the learning of how to succeed electronically and manage these channel conflicts.

## QVC

QVC (Quality Value Convenience) is *the* TV home shopping broadcaster. Established in 1986, it has demonstrated the extraordinary reach and potential of TV home shopping. It reaches into 54 million homes across the USA – more than half of all US households – and currently receives more than 70 million phone calls a year for products ranging from jewellery, household and hardware through to health and beauty and clothing fashion brands. In 1996, it shipped around 50 million packages in the USA.

QVC opened in the UK in 1993 and took time to establish itself. Its breakthrough came in 1995 when it negotiated with the British Satellite Broadcasting company (Rupert Murdoch's satellite venture) to be an integral part of that company's multimedia channels package. As a result in the UK it was able to get the access to consumers it had previously been denied. It now reaches into more than 5 million homes (25% of all households), has a customer base of over 1 million and is seeing sales grow at 30% plus a year.

As with First Direct, QVC has worked hard to make home shopping an easy, reliable and pleasurable experience. It has developed a sophisticated telecommunications network able to handle 50,000 purchases an hour and offers a 24-hour next-day delivery service. It ensures the product quality meets consumer needs, makes the shopping experience enjoyable and offers full guarantees and refunds. It looks to provide all the key ingredients converting uncertain consumers into becoming TV shoppers and paving the way for others to develop their own successful TV shopping operations.

*Dell*

Dell are pioneers in telemarketing and selling. They have broken the mould in their industry sector by consistently and persistently selling computers direct to the consumer, bypassing the traditional network of electrical wholesalers and retailers. In 1997 Dell's turnover reached $7.8 billion, up 47% from the previous year and with sales continuing to grow each month at high rates.

Dell started out twelve years previously confounding industry experts who claimed it would be impossible to sell something as complex as a computer over the phone. But Dell's argument was that everyone had a phone, many sat next to one all day long in their offices, it could provide immediate connection not just for selling but to handle queries, provide after-sales service and 'keep the customers satisfied'. Many industry commentators also believed no consumer would provide their credit card number over the phone for security reasons. Again, Dell showed that this was not the case once consumers could believe in the company and trust its professionalism.

In that regard Dell put its faith in its front-line employees and saw them as the source of its competitive advantage. It has made enormous efforts to ensure they are educated about their products, can easily access and assess the technical specification data and interact efficiently with customers: 'We will use our employees to deliver the personal touch that many customers desire – you don't need physical contact to do that.'

Dell not unexpectedly was one of the first companies also to pioneer marketing and selling over the Internet and does an increasing part of its business on-line. It's vision is to have all its customers eventually do business through this medium, but that vision assumes that phone and video facilities will be integrated to ensure communication can remain personal and interactive.

As it develops its Internet business Dell is also experiencing its own form of channel conflict with its existing salesforce and telemarketing operation querying whether their jobs and customer relations are under threat. But the company's view is that the Internet simply enhances the total customer communication package and that if less time if required for selling or information-giving over the phone then more time is available for customer account management, relationship building and after-sales service.

*Levi Strauss*

Levi Strauss has long been one of America's most admired corporations and has made its name in the world of blue jeans and casual clothing. But as a traditional manufacturing company, Levi's often felt too removed from its end-consumer and felt frustrated that its consumer connections only got as far as its factory gate.

Levi's has shown what manufacturers of consumer goods can achieve by pushing out into the consumer arena, opening its own stores and working hard to build stronger customer relations. It is this greater proximity to its market place that helped Levi's realise its customers still weren't satisfied. They wanted still better service, better fit, better value. In a world where many products are perceived as commodity-like, Levi's consumers expected something extra from the market leader.

As one response to this challenge Levi's have made their whole sales, marketing and distribution much more electronic. For its distributors it has developed an interactive operating system that automatically allows reordering and enables other retailers to tap into Levi's own internal manufacturing processes, track where the order is and even alter order configurations within certain guidelines. (Fedex and UPS are other leading-edge examples in this.) Distributors as a result feel they have much better relationships and are more in partnership with Levi's.

In Levi's own stores, the company has introduced a simple interactive kiosk which enables the store assistant or customers themselves to input their personal measurements and get a 'Custom Fit' pair of jeans made up and ready for collection within a couple of weeks. Lead times for this service are already reducing and a one-week turnaround is already being targeted.

Levi's are now considering the possibilities of offering this same complete service over the Internet so consumers can shop with even greater confidence from home. After all, for many of Levi's consumers buying a pair of jeans is simply a replacement for a worn-out pair, there is no need to visit the store to try them on and a measurement facility can be easily offered on-line.

## Amazon.com

This virtual bookseller ranks among the most famous sites on the Internet. Founded as recently as 1995 it has set the pace in electronic shopping generating more than $40 million of sales in the first six months of 1997. It has created a mini revolution in its industry sector forcing once sleepy chains like Barnes & Noble to wake up to the realities of the electronic age.

As a pioneer Amazon.com has enjoyed enormous publicity, helping to create awareness and also quickly establishing its brand name among the book-buying public. Jeff Bezos, the company's founder, was convinced from the outset that consumers would enjoy the experience of browsing books electronically, reading reviews and excerpts from a particular book and being able to have it delivered within a day or two. For Bezos it wasn't just about providing a convenient service, it was also about being able to offer a one-stop shop making available the widest possible selection of titles – 'if it's in print, it's in stock'.

Amazon.com's early trading saw significant losses and when Barnes & Noble and Borders, the established book retail giants in the USA, woke up to the electronic opportunities, many believed Amazon.com would simply be muscled out. But Bezos has stuck to his guns, milked the TV and press publicity about his site, marketed and priced aggressively and is emerging as the dominant on-line bookseller.

In fact bookselling over the Internet is now one of the more sophisticated industry sectors to have developed in this way. Market commentators have moved on from debating whether virtual book shopping will survive to examining the next stage of its development. Competition between booksellers on the Internet is intensifying. We've already seen our first Internet pricing war in summer 1997 with companies outdoing each other to offer greater discounts. But on a more long-term basis, issues of branding, service levels and tying up on-line distribution deals will become more prominent. For example, Amazon has secured deals with America On-Line and search engines Yahoo! and Excite to guarantee on-line traffic and ensure high prominence.

What Amazon so admirably demonstrates is how quickly a business can develop and grow in the Internet environment. In Amazon's case from $500,000 of sales in 1995 to a forecast $80 million in 1997 and predicting about $400 million for 1998! It also shows how easy it can be in this new electronic age to bypass established physical retailers and reach directly to the consumer. The key, as demonstrated by Bezos, is the conviction and belief in the potential of this way of buying and selling and the pioneering determination to make it succeed. Right now this 'pioneering spirit' can be seen as the domain of the romantic and the adventurer. How long before it becomes essential for survival?

*Tesco*

Tesco is Britain's largest food retailer and has been leading the trial and experimentation of electronic shopping in its sector. Its web site is one of the most frequently visited by the UK consumer base and it has been steadily building sales.

Its latest moves in 1998 include the extension of its home shopping service throughout London. Its Internet superstore offers customers the ability to purchase any of the 20,000 product lines typically available in its stores. Tesco mails all would-be electronic shoppers a CD Rom with the shopping list and orders are made on a return path on-line via CompuServe. Each month in addition Tesco mails consumers a full-colour catalog detailing special offers and highlighting particular product categories: 'We're constantly looking to improve our offering and we have to keep thinking about the future and what our customers will demand.' Next-day delivery is the usual arrangement but shoppers can also order goods up to 28 days in advance. Deliveries are made by a sub-contractor with a fleet of refrigerated vans.

Tesco acknowledges it still has much to learn about Internet shopping and home delivery and that the service levels and interactions that take place are a long way from their potential: 'We have a history of innovation and we're learning on this one at every step . . . it's important to be at the forefront as the Internet develops . . . there would be nothing worse than having to play catch up.'

Tesco certainly still sees its physical real estate as its principal distribution channel and while it is treating Internet opportunities more seriously than many of its rivals it's a long way still from seeing that as its future direction. What Tesco is already experiencing though is growing levels of interest on the back of the publicity for its home-shopping service. It is also currently enjoying getting new customers who cannot get the same electronic shopping service from their own regular supermarket.

Tesco is actively putting itself in a prime position to benefit from increasing electronic shopping interest. Most importantly, it is learning how to make that route to the consumer work, is establishing itself in the consumers' eyes as being the leading edge exponent and will thus be best able of all its competitors to effectively respond and see through the inevitable evolution of its existing retail estate.

**General trial and experimentation**

Aside from the major pioneers there is a wealth of trial and experimentation taking place across the whole retail shopping scene. This can be categorised by looking separately at what steps different players in the distribution chain are taking and how they see their interest in electronic commerce. There are many case examples with considerable innovation taking place. We can consider:

- Manufacturers
- Retailers
- New intermediaries
- New entrants
- Consumer clubs.

*Manufacturers*

Manufacturer supplier companies are considered in more detail in Chapters 12 and 13 but to date they have mostly kept a low profile. They are concerned that they may disturb existing channels of distribution and undermine currently important customer relationships. This has dampened their enthusiasm and caused them to downplay any initiatives. Only a handful of companies have announced their plans:

- Unilever is looking to develop its own direct links with its end consumers:

    'We're looking at diversification into new service-related methods of channelling our food and personal products to the consumer. Because of fundamental changes in the retail market we are seeing the emergence of alternative channels of distribution . . . unless we are careful we will find ourselves boxed in to what is sold through a consolidating block of retailers.'
    (*Sunday Times interview with Niall Fitzgerald, Chairman of Unilever*)

- All the major car manufacturers from Audi to Volvo have set up web sites and are experiencing tremendous interest. One specific example is Rover Cars, the BMW subsidiary, who have been experimenting with selling over the Internet. The company has set up a point and click facility to demonstrate the features of the cars and includes full motion video sound and graphics which can change colour and specification. There is also a price list based on various finance deals. The system is used particularly by the dealers as a high-speed ordering link, demonstrating cars not in the showroom and expanding the range of information available. Consumers can access this site on-line but lack of sufficient bandwidth in many home PCs means downloading times are still slow and picture quality not yet up to acceptable levels. Rover nevertheless say they are committed to strengthening this way of doing business.

- Guinness and Allied Domecq are both examples of manufacturer brewers who have set up on-line. While their sites are relatively well developed they are still at the stage of providing information to consumers and haven't yet progressed to proactive electronic selling. They are still exploring how to develop electronically without upsetting existing distribution channels. Guinness's site is nevertheless so popular that many have downloaded their on-line advert for the product as a screen saver. The site also sells related merchandise like sweatshirts and has received orders from all over the world.

*Retailers*

This group will also be considered in more detail in a later chapter (Chapter 6) but while Tesco has been taking the lead, other retail groups from grocery to banking to betting shops have also been trialling new approaches:

- Sainsbury, the major UK rival to Tesco's food stores, has established an 'Order & Collect' operation. This involves customers ordering by phone or fax from a catalog, having the goods picked for them in store and then coming to collect later in the day or at some other prearranged time. The service costs an extra £2 and customers need to sign up to take advantage. It's done on a local store basis and there are presently limits on the number of 'order and collect' customers per store. Sainsbury will also help customers develop their own shopping lists. Appointments are made with customer service representatives who will then spend around 2 hours walking with the customer around the store compiling a standard list of basic shopping items which can then be ordered by simply quoting a reference number over the phone.

Sainsbury also have an Internet/home delivery option and are in talks with British Interactive Broadcasting for a store offer on the new digital TV shopping platforms to be launched in 1998 (see Chapter 9).
- Waitrose, another UK grocery chain, has launched an Intranet shopping scheme with ICL Computers. The initiative will allow the 500 ICL staff to sign up, order their shopping from their desks in their office and pick up from the office lobby area on the way home.
- Lloyds TSB, the UK retail banking group, opened 1998 with the announcement it was setting up a 24-hour telephone banking service to rival First Direct: 'We want to move with the times and growing number of our customers are looking for this kind of service.'
- Barclays Bank relaunched its web site in 1997. Its first effort set up in 1995 was 'an important experimental test bid' but did little business. Its relaunch involved a fundamental reworking and an easier-to-use format, 'We have seen stunning growth during 1997 with over one and a half million visitors and we are looking for the J curve of performance as we come out the learning curve.' The site itself is still seen as an incremental growth opportunity and Barclays will not yet acknowledge that it will have any meaningful impact on their branch business.
- Major UK betting shops like Ladbroke's and Coral's are trialling new web sites and are considering offering trackside betting on horse races all over the world from the comfort of the home. Ladbroke's for one has announced that it is set on establishing the on-line operation but is still working out how to ensure the betting and payment can be kept secure and confidential and there is an effective shield preventing children under the legal age from betting. The security and confidentiality side they expect to have resolved at latest by end-1998 and they hope to prevent under-age gambling through a mix of personal subscription and credit vetting.

*New intermediaries*

The arrival of electronic shopping has created opportunities for new companies to come in and capture a part of the restructuring value chain. This is particularly in the area of home delivery. Existing manufacturers and retailers don't have such an infrastructure in place and typically sub-contract transportation and warehousing. This has facilitated new dedicated home delivery companies to enter the market place and start taking over the previous role of the retailer as distributor closest to the end-consumer:

- Peapod is an on-line grocery shopping and delivery service. its aim is to 'relieve Americans of the time and trouble of buying groceries'. They are looking to capture just a small share initially of the $400 billion US grocery market but 'even a 1% share in the next two or three years would turn us into a substantial grocery distributor especially if concentrated in our five target regions'.

  Peapod takes order by phone, fax or computer and delivers to the door. Their Vice President Thomas Parkinson says they want to change the way people shop. Peapod are encouraged in their ambitions by independent research organisation Forrester whose early forecast saw a growing rise in home delivered food and drink sales (Figure 3.1). Peapod have already found that barriers to buying fresh produce can be overcome and its top five selling product lines include fresh fruit and vegetables. Vice President Parkinson has said that Peapod is rapidly proving its reliability on product quality, 'shoppers are flocking to us'. Future development has now been further underpinned by the arrival of a 20% shareholder in WPP, the global marketing services group, who can bring funding and advertising muscle to complement Peapod's activities.

**Figure 3.1** Projected Food and Drink Sales on the Internet

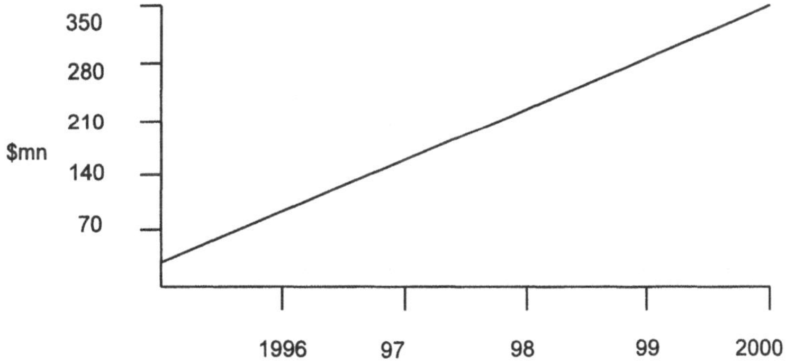

*Source*: Forrester Research (1996).

- Shoplink is another venture looking to attack traditional distribution channels and facilitate new ways for customers to shop. While Peapod are working with a range of different stores like Wal-Mart or Kroger, Shoplink has set up greenfield. Started in 1997 the company aims to establish its own refrigerated warehouses to service the eight major cities across the USA, starting in Boston. Bankrolled by parent Elcom

International, CEO Robert J. Crowell is not surprisingly an enthusiast:

> 'Our aim is to take the grocery stores out of the equation. We will receive deliveries direct from the manufacturers with our own onward home delivery operation. We can save on all those retail real estate costs and can therefore afford to deliver to individual homes at the consumer's convenience for only a small delivery charge.'

- Streamline, another US operator set up along similar lines, has announced a target of 1 million households buying grocery direct through their operation by 2005.
- In the UK, similar dedicated home-shopping intermediaries are busily establishing themselves. Flanagans works presently throughout the western side of London and has links with Sainsbury, Food Ferry is more like Shoplink, targeting certain areas but buying in product into its own warehouse. The Cotswold Shopping Service provides an example of a neighbourhood home-shopping service, buying goods on order from any store and organising their home delivery.

*New entrants*

Amazon.com have paved the way but others are following hard in its footsteps setting up as dedicated on-line providers:

- During 1997 a plethora of new sites were set up in the Financial Services field. New companies are providing information and transaction services so that it is possible to deal in company shares on the screen but also to buy motor insurance or shop around for a mortgage. Such services can be cheap both to set up and run. Indeed there are so many sites being established that the regulating authorities are struggling to control those enterprises that appear to be operating in their jurisdiction.
- Music sales are another rapidly growing Internet category. Research consultancy Jupiter Communication forecasts on-line music sales in America to be worth $1.6 billion by 2001 – that is 13% of the US music market. New companies like CDNow provide information, band details, discographies and track listings. There are audio samples for many different albums and the whole web site is easily laid out enabling would-be shoppers to follow the categories of music they are

most interested in. Pricing in the USA is the same or cheaper than in-store and delivery can be within 48 hours. While CDNow.com is available in the UK, the UK specialist site CDDirect.co.uk is significantly more accessible for local consumers.
- New sites offering travel and holidays means there may be no need to deal through a travel agent and certainly no need to visit their shops. Most of the major airlines, for example, have their own web sites with clear on-line booking facilities. All the major car rental companies and large hotel chains now have excellent arrangements to book direct with ever-better information and display on the product and services being offered. For holidays, a number of destinations e.g. Skiing in Vail, have come together to set up their own web sites that can then take the consumer into general resort information or individual hotel room availability.

*Consumer Clubs*

This is a new phenomenon and the late 1990s has seen the emergence of only a few examples. They are discussed later in Chapter 7. But they are likely to represent a growing force in the distribution chain. Once consumers begin to become comfortable with the Internet as a communication and transaction device they'll begin to learn not just how to use it to get information and buy goods, they'll also begin to see it as a means of taking greater control for themselves for what they buy, when they buy and the price they buy at. They will no longer be dependent on the few stores they can reasonably drive to in their neighbourhood. As this consumer awareness grows, so the Internet will enable like-minded consumers to 'club together'. Some of this clubbing will be for social purposes, some of it for lobbying or political ends, but some element of it will be to get better buying terms from manufacturers or distributors. A 'mother and baby club' is an obvious example where mothers in a neighbourhood get together, organise collectively their total weekly diaper requirements for example from Procter & Gamble and order direct from the manufacturer themselves, expecting next-day delivery and on better terms than available from distributors.

These 'buying groups' will develop through the next decade and we will likely see a few pioneers spotting this as a new business opportunity and being at the forefront in encouraging and exploiting it.

\*  \*  \*

There is certainly a vast amount of electronic shopping activity. Some of the more notable examples have been described but it's worth also referring to some of the more broadscale initiatives. These aren't limited in their impact. They affect all parts of the distribution chain and so challenge not only retailers but also manufacturers and other existing distributors.

One example can be found in Palo Alto, California. There, 1997 saw an initiative to wire up the whole city with full two-way Internet capability as part of a 'Smart Valley' project. The aim is to establish an electronic community, linked by high-speed networks, able to communicate and conduct commerce with ease and reliability. Project director is the Palo Alto mayor Liz Kniss who has commented how this investment will 'open up possibilities and potential we haven't even imagined'.

Similar projects have been established on a smaller scale in France and in Spain. Finland has capitalised on its already sophisticated telecomm infrastructure to accelerate Internet connections and activity. A few streets in London have been converted by way of experiment, in one case sponsored by Microsoft, to test out consumer interest and responsiveness.

As these community-wide initiatives take root so electronic sellers will be encouraged to increase their own investments and so it will be easier still for consumers to buy what they want. There's a virtuous circle that will gradually take shape and solidify (Figure 3.2).

**Figure 3.2** Electronic Shopping: The Virtuous Circle

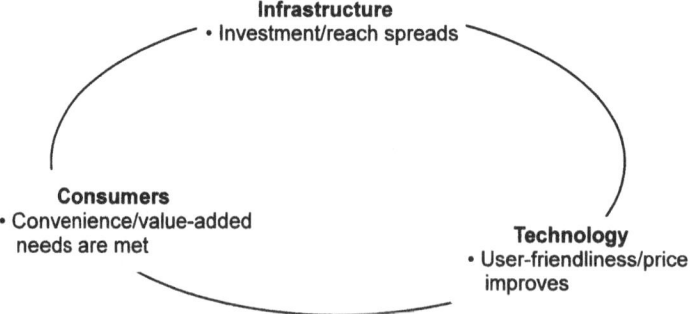

# 4 Future Growth of Electronic Shopping: Look Out for 2005!

There's so much activity and investment going on, but when will it reach critical mass, when will it truly begin to shape mass consumer shopping habits? All the evidence is pointing to around the middle of the next decade, to c.2005. By that stage it looks as though there will be sufficient infrastructure and technology in place, experimentation and improvement will have reached a point of maturity – though still far from complete – and consumer purchasing will start proving notably easier, more enjoyable and more rewarding.

In this chapter, the evidence pointing to 2005 is examined, describing some of the key dynamics bringing this critical mass date so near in time. For the reader searching for more detail, Chapter 9 drills down further, looking at both the impressive range of technology and infrastructure options that are being developed and at how consumers are already impatient for these kind of innovations.

As we begin to examine the shape of things to come, let's first envision what the brave new world of electronic shopping might be like. How will life-styles and habits change, how will it save time and improve the quality of life? Here are a few possible scenarios:

- It's Friday, it's 6.30pm and Deborah is just finishing work. Up till the end of last year Deborah used to have to rush to the supermarket to do her weekly shopping. She likes to keep her weekends free but spending Friday evenings after a hard day's work pushing her shopping cart up and down the crowded aisles was a bit of a nightmare. Recently, however, it's all changed. She can leave the office on Fridays and head straight out with her friends. What's happened? Simply that her local supermarket has at last expanded its home-shopping service to her area. They've got her regular list of basics and each Friday she can just phone, fax or e-mail them, quoting her reference number, and it automatically triggers the reorder. There's also the opportunity to add on the specials she wants that week – fresh strawberries and cream because it's summer and

Wimbledon. While she's now out relaxing and enjoying herself someone else is walking round the food warehouse picking her order. It's then all delivered to her, as arranged, freshly picked and packed on Saturday at 10.00am just as she's finishing breakfast. Next year her supermarket have announced they'll be providing her with special food-supply bins for free. They'll be fixed in a suitable place to the side of her house and there will be a digital lock on the lid. The supermarket delivery van can then deliver whether she's at home or not. Only they will have the digital key code and the food will be packed in special chilled containers keeping things fresh.

- John has never liked shopping much. He always felt a bit embarrassed standing in the middle of a shop trying on new clothes and shoes while passers by ogled and no doubt thought his choices ill-suited and inappropriate. To some extent, those days are gone. For example John has just discovered the new Custom Shoe Shop. It's in that great big shopping centre that's opened. You can drive there and get just about anything. In the new shoe shop, John is shown into a personal cubicle and invited to sit down in front of a computer screen. The instructions tell him to place his feet on special plates. He touches 'start' and his feet are photographed from a variety of angles and detailed measurements taken. There's a new global standard issued by the Shoe Manufacturers' Association and John is pleased to see they've approved this particular equipment. While he's waiting a shop assistant has brought him some coffee and croissants and the daily newspaper. It's all so civilised. John has seen some of the Custom Shoe Shop's adverts but he now goes back to the screen and scrolls through the range of shoes and accessories. He can put any buckle or bow on any shoe in any colour. If he needs help the assistant is on call. But it's all very easy. He chooses a pair he likes and the computer shows pictures of what they would look like on and he can look at different colour suits and trousers to check they match. Now he has the choice of a 2-hour express service, he can go and check out the rest of the mall and come back and collect or he can have them delivered next day. Because it's John's first time he says he'll come back later. When he does the shoes fit very comfortably. Now they have his details and preferences, next time he needs a pair of shoes he could access their web site and do all this from home or the office.

- Peter has just received an e-mail telling him the interest rate on his mortgage is going up next month. Peter is concerned and is sure he can get a better rate. He goes on-line and uses one of the search engines to look at different mortgage rates that could be available and to find the one that offers him the best value. The search engine looks all over the world and identifies one bank in San Francisco and another in London that might have what he wants. Peter fills in a basic form on-line setting out details of how much he wants to borrow, the equity in his house etc. and the search engine automatically notifies the two banks, informs the banks that their offers are being matched and compared and asks each for its best counter offer. The London bank seems more flexible and starts setting out terms that Peter likes. It will be cheaper than his current mortgage and there will be no switching costs. The bank makes Peter an offer subject to the usual conditions and Peter e-mails his financial advisor to send through the necessary documentation to the bank. The whole process took just half an hour and Peter can't help but remember the first time he went to get a mortgage, had to wait in the bank to see the manager, how long it took, how many banks and brokers he had to visit and call. He can now get on with enjoying his weekend.

These scenarios are no longer stories from science fiction. The technology exists and suppliers are advanced in experimenting with it. In 1998, at the time of writing this book, it is already possible to shop from home, have clothes and shoes measured automatically (look at Levi's 'custom jeans') and search the Internet to compare prices. Given the rapid pace of development these services can only improve in speed, quality and price. In fact there is a further evolution already being developed in the lab, though further off in terms of its mass commercialisation. It's going to be yet a further step forward, making electronic commerce not just more convenient but also exciting and fun. It's 3-D virtual reality and while it hasn't yet impacted the world of shopping it's already on many Internet companies' horizons. We can get a view of how 3-D will work with this extract from Michael Ridpath's *Trading Reality*. With 3-D graphics, consumers can use their extensive knowledge of the physical market place to shop for products in the virtual store:

'Most of the break-throughs in computer technology over the last twenty years have come from getting more and more power on smaller and smaller chips . . . But you remember I spoke about Graphical User Interface, how people will talk to computers? That's where the next strides will come. And virtual reality is the ultimate user interface.

When a person is actually in a computer-generated world, can talk to it, point within it, then the computer disappears as a barrier. Computer and user become one. I'm convinced that a whole new range of human activity will become possible. Anything to do with communicating between people in different physical locations will be vastly improved.

You will need a computer with special software to run the virtual world. Then a head-mounted display with two little screens, one for each eye, earphones to give sound in stereo and a sensor that tracks where you're looking. So when you turn to the left the image you see changes to the view to your left in the virtual world. Special gloves can recreate a sense of touch. With the senses of sight, sound and touch all totally immersed in the computer-generated virtual world you really are in virtual reality.'

We can see this future world rushing at us, about to change our experiences and perspective and expectations of what can be done and what can be achieved. To some extent we can't imagine it, yet ideas and images come from movies we've seen, books we've read, news features on the latest computer product innovations. What's so extraordinary is that it's now coming alive – it's the fast-approaching reality.

One of the major forces driving its reach and penetration into our lives is the rapid spread of the Internet and the number of people that can access it. It's coming into homes through both the PC and the TV. It's coming over telephone wires, cable and satellite. It's available in most offices and often at each person's desk through their own PC. It's available while people are travelling at airports and at kiosks. It reaches across the world, its messages are universal, its products and services the same. There are no barriers. Payments can be in local currency, software can be used to translate an English-language web site into the local dialect, everyone can in principle get everything. Of course, there are social and cultural hurdles. Some countries and people would never be able to afford the equipment but there's no doubt that across the western hemisphere at least commerce is going global on the Internet.

A look at some of the statistics provides compelling evidence for how quickly things are expanding. We can consider in turn:

- infrastructure development
- size of the Internet economy
- who's using the net
- consumer acceptance
- country snapshots

## Infrastructure development

There are several dimensions here:

- The number of host computers on the Net (the computers that deliver information and services to users) has grown from 100,000 in 1990 to 10 million by 1995 and is projected to grow to more than 100 million by just the end of the decade
- The number of commercial web sites grew from 350 in 1994 to 220,000 in 1996, estimates for the year 2000 are for more than 1 million
- In terms of the number of households on-line, Table 4.1 shows the extraordinary growth and penetration in such a short amount of time
- Another way of seeing this household penetration is from research in Table 4.2, showing the growing percentage of households with PCs and in Table 4.3, the growing number of PCs that are wired to the Internet.

Table 4.1  On-Line Households, 1996 and 2000

| Country | 1996 (Actual mn) | 2000 (Forecast mn) |
| --- | --- | --- |
| USA | 15.4 | 38.2 |
| Europe | 3.7 | 16.5 |
| Asia | 3.4 | 10.0 |
| Japan | 2.6 | 7.1 |
| Australia | 0.5 | 2.0 |
| Germany | 2.0 | 6.9 |
| UK | 0.7 | 4.3 |
| France (excl. Minitel) | 0.2 | 1.2 |
| Scandinavia | 0.5 | 2.0 |

*Source*: Jupiter Communications.

Table 4.2  Growth of European* Households with PCs, 1994–2000

| Year | % households |
| --- | --- |
| 1994 (actual) | 17 |
| 1995 (actual) | 22 |
| 1998 | 30 |
| 2000 | 40 |

Note:* Western Europe only.
*Source*: The European Information Technology Observatory.

Table 4.3 Penetration of US Households with Computers and Internet Connection, 1995 and 2000

|  | 1995 (%) | 2000 (%) |
| --- | --- | --- |
| Home PCs with Internet | 21 | 43 |

Source: The Internet Society.

As a specific example of infrastructure development, Verdict research has set out (Table 4.4) its view on how the UK especially will become wired up to the Internet. They foresee higher levels of penetration and usage and forecast that more than one-third of the UK population will be on-line by the year 2001.

Table 4.4 UK Growth in Internet Connections, 1997–2001

| Year | Home Internet a/c holders (mn) | Growth (%) | Corporate Internet (mn) | Growth (%) | Satellite digital TV (mn) | Growth (%) | Digital terrestrial TV (mn) | Growth (%) | Total (mn) |
| --- | --- | --- | --- | --- | --- | --- | --- | --- | --- |
| 1997 | 1.55 | – | 0.8 | – | – | – | – | – | 2.35 |
| 1998 | 3.11 | 100% | 1.9 | 150% | 0.04 | 20% | 0.01 | 35% | 5.2 |
| 1999 | 5.45 | 75% | 4.6 | 130% | 0.09 | 125% | 0.01 | 150% | 10.2 |
| 2000 | 6.81 | 25% | 9.1 | 100% | 0.27 | 200% | 0.02 | 200% | 16.4 |
| 2001 | 7.49 | 10% | 13.7 | 50% | 0.88 | 225% | 0.90 | 400% | 23.0 |

Source: Verdict.

A sophisticated infrastructure is gradually taking shape around the Internet which will facilitate and accelerate the increasing levels of electronic commercial activity that are predicted. Rapid PC penetration is combining with fast cable and improving telephone wire links to enable quick net access and the downloading of richer material with higher video and visual content. Developments with digital and satellite broadcasting promise to bring net access to millions more households, avoiding the expensive laying of fibre optic cable. The emergence of open standards in development tools and at the network protocol level has made it cost efficient for software developers to create interoperable products. Payment security systems are now enabling web site operators to make data transmissions on credit cards safe. Data encryption software is improving and spreading providing further reassurance on confidentiality and privacy.

There is a growing maturity following the early days of hype and exploration and there is an impressive gathering of companies providing hardware, software, telecommunications, media and content racing to provide an attractive buying and selling environment for the twenty-first century.

**Size of the Internet economy**

Total commerce over the Internet – both business-to-business and business-to-consumer – is forecast to reach $200 billion by the year 2000. In the USA alone it is expected that Internet revenues will reach $66 billion by the end of the century, setting the pace that other countries will doubtless follow.

These revenue forecasts are drawn from an extensive series of interviews with over 40 CEOs whose companies do business on the Internet today. They highlight five major industry segments with the biggest on-line revenue potential: infrastructure, access, content, financial services and consumer retail. These CEOs say the key drivers of electronic commerce are the number of households and businesses going on-line and the increasing availability of effective technology solutions from the vendor community.

Spotlighting business-to-consumer Internet revenues, most forecasters see $10–$12bn of revenues by the year 2000 (Figure 4.1, with a recent *Business Week* survey reinforcing views that the top end of this range will

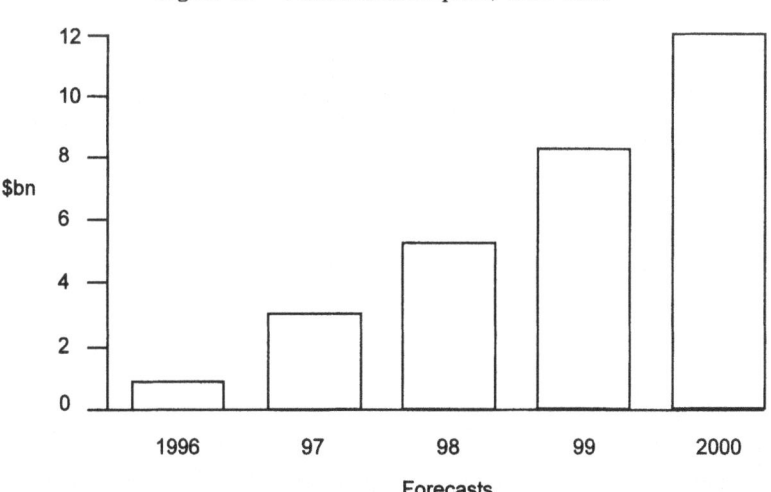

**Figure 4.1** Total Internet Spend, 1996–2000

*Source*: *Business Week*/Forrester. (The most recent forecast is $17 bn by 2000.)

easily be reached. These figures exclude catalogue home-shopping sales and other established phone/fax operations. In 1997 those longer-established forms of electronic shopping accounted for nearly $80 billion of sales, just taking into account North America.

The same *Business Week* survey (Table 4.5) named the top ten selling product and service areas.

Latest reports at the start of 1998 suggest further headlong developments in Internet sales, with *Business Week* finding that growth is coming in every product category – from 'airline tickets to tennis rackets'. American On-Line President Robert Pittman has commented that consumers are getting increasingly familiar with the idea of the Internet and beginning to learn they can shop satisfactorily through it: 'people are beginning to move from window shoppers to buyers . . . few companies of any size haven't now got some kind of web site'.

**Table 4.5** Top Ten Internet Shopping Categories

1. PC hardware
2. Travel
3. Entertainment
4. Books and music
5. Gifts and flowers
6. Apparel and footwear
7. Food and beverages
8. Jewellery
9. Sporting goods
10. Consumer electronics

*Source*: *Business Week*/Forrester.

**Who's using**

Traditional home shopping by catalogue mail-order or by phone and fax has always been especially popular among women who are the majority users. The Internet, however, was originally the preserve of men and in its early days had a very skewed demographic roughly equating to the computer nerd. Times are changing though and the latest surveys on who's using the Internet show a much more representative balance emerging with people of all ages, backgrounds and of both genders.

GVU, a US-based research organisation, has carried out widely read surveys on Internet usage and the seventh was published in the summer of 1997. Among other things this shows that 33% of Internet users are now women. This proportion rises to c.40% in the USA. In Europe it is around 20% on average but growing since the previous year's research.

Other studies by NUA and Lexmark International show that women are increasingly using the PC for a variety of applications ranging from planning the family's finances, running educational programs, e-mail, newsgroups and bulletin boards to a growing interest in shopping facilities. Researchers predict that increasing comfort and experience of using PCs among all members of the population will lead to greater investigation of Internet options and thereby more trial of what can be purchased. This development is now being fuelled more particularly by a number of print magazines, some targeted specifically at women and featuring web home pages and bringing attention to what's available. In addition there are a wider number of specialist web sites such as Hearst's HomeArts Network, Women's Wire and Village Parent Soup which are targeting new women readers.

For would-be Internet providers other demographics also look appealing with good acceptance across most age ranges and social/income groups (Tables 4.6 and 4.7).

**Table 4.6** Internet Electronic Shopping, by Age

| Age group* | (%) |
|---|---|
| 15–24 | 34 |
| 25–34 | 26 |
| 35–44 | 25 |
| 45–54 | 18 |
| 55–64 | 14 |
| 65 + | 7 |

Note: *Average age is estimated to be 34.
*Source*: GVU, FT Management surveys, Healey and Baker.

**Table 4.7** Internet Electronic Shopping, by Socioeconomic Groups

| Social group | (%) |
|---|---|
| AB | 33 |
| C1 | 20 |
| C2 | 17 |
| DE | 16 |
| All adults | 21 |

*Source*: GVU, FT Management surveys, Healey and Baker.

Internet users are spread all over the world but the energy and leadership is firmly rooted in the USA. This is consistent with US corporate attitude to IT investment generally which has traditionally been significantly more progressive and proactive than all other major trading nations (Figure 4.2).

**Figure 4.2** Average IT Spend per Worker per Year ($)

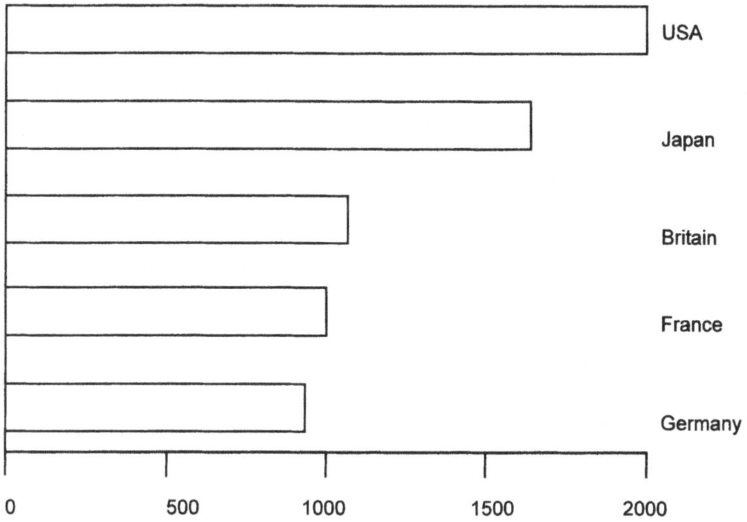

|  |  |
|---|---|
| USA | ~2000 |
| Japan | ~1600 |
| Britain | ~1050 |
| France | ~1000 |
| Germany | ~900 |

*Source: Business Week*/IBM.

Such is the lead the USA is establishing that some commentators have expressed concern that other leading western countries are in danger of falling behind. More recent initiatives in both the UK and France, for example, do however show a heightened awareness of the investment gap and plans are being put in place to catch up (see 'Country Snapshots' on p. 59).

**Consumer acceptance**

Some 60% of adult consumers have shopped remotely and satisfactorily from home. There is a comfort and confidence in using established devices such as phone, fax or the mail. It is widespread and part of the shopping scene. Familiarity with many products and services makes it easy to buy without having to visit the store again. In principle any improvement to this home-shopping experience should be appealing. This

## Future Growth of Electronic Shopping

is especially so if it can enrich the quality of interaction and information communicated. In this regard no one doubts the potential of the Internet to fulfil that goal, but all recognise the technology has some way to go. A key question in this book is when will it reach a sufficient level of maturity and capability that a critical mass of consumers will become engaged and involved in it on a regular shopping basis.

Consumer interest in the Internet is already high and it appears there is a strong platform on which would-be Internet sellers can build. Tables 4.8 and 4.9 show there is an established level of confidence in general use of new technologies and significant numbers of people prepared to consider different forms of remote shopping.

**Table 4.8** Confidence in Using New Technologies, by Age Group

| Age group | % confidence |
|---|---|
| 16–24 | 51 |
| 25–34 | 40 |
| 35–44 | 38 |
| 45–59 | 26 |
| 60+ | 11 |

*Source*: FT Management report; Henley Centre.

**Table 4.9** Proportion of People who would Consider Remote Shopping

| Shopping method | % of respondents |
|---|---|
| Postal system/mail-order | 50 |
| Telephone | 34 |
| Cashpoint | 25 |
| Fax | 25 |
| TV | 24 |
| PC Internet | 15 |
| PC CD Rom | 14 |
| Electronic kiosk | 9 |

*Source*: FT Management report, BMRB.

A recent study by Motorola in association with the *Wall Street Journal* found that this platform of consumer confidence applied to a wide variety of electronic goods and services (Table 4.10).

Among certain target segments, acceptance of remote shopping is especially high. A recent survey among 25–45 year-old males suggests that this segment already represents a highly attractive marketing target for electronic sellers (Table 4.11).

**Table 4.10** Comfort in Conducting Services by Computer

| Service | % respondents |
|---|---|
| Ticket/Travel reservations | 63 |
| Home Banking | 51 |
| Education/Training | 42 |
| Shopping | 32 |

*Source*: Motorola.

**Table 4.11** Attitudes to Electronic Commerce

| Question | % respondents answering Yes |
|---|---|
| Ever bought via Internet, multimedia kiosk or interactive TV? | 33 |
| Ever ordered and paid over the telephone? | 98 |
| Would pay a premium to have goods delivered to the home? | 78 |

*Source*: IMRG.

These are remarkably positive levels of interest and acceptance among consumers at large and in certain segments in particular, all the more so because electronic commerce is still in its relative early days and still grappling with inevitable early teething troubles. It appears that many see Internet shopping as simply a natural evolution from more traditional phone/fax/mail-ordering. Comfort with the PC, familiarity with brands, readiness to rely on suppliers and providers with a solid reputation all leads to a quiet comfort with the new medium.

Business at large has naturally also observed these same consumer trends and indications and has a growing awareness of their long-term potential. For example, a *Forbes* magazine survey of chief executives in 500 top US companies found 40% of executives believing that 40% of all transactions would take place over the Internet in the future. They also saw consumers becoming more sophisticated shoppers using the Internet first, to obtain product information, then to compare prices and then interactively consult company sales executives before then ordering – a series of steps which the Internet readily facilitates.

These predictions among company executives on the impact of the Internet are spread wide across all sectors of industry. For example, an IDC research study asking how many felt electronic commerce would be 'critical to very important' in the coming years generated strong

*Future Growth of Electronic Shopping*

**Table 4.12** Importance of Electronic Commerce

| Sector | % respondents saying High Importance |
|---|---|
| Retail | 32 |
| Banking | 29 |
| Insurance | 32 |
| Process manufacturing | 26 |
| Transport | 29 |
| Healthcare | 14 |
| Business Services | 43 |

*Source*: IDC.

responses with executives in the business services arena especially showing an acute consciousness of the shape of things to come (Table 4.12).

**Country snapshots**

With so much of the Internet development taking place in North America and in the English language it provides a useful perspective and contrast to examine two of Europe's non English-speaking powerhouses – France and Germany. We can also take a snapshot on Singapore, showing the remarkable strides being made in one of Asia's leading economies.

*France*

France was the first country in the world to get wired up. During the early 1980s, the French government through the state-owned French Telecommunications company issued to millions of households for free a 'Minitel'. A small simple low-power computer terminal, its original purpose was to give French consumers direct access to an electronic phonebook with the Yellow Pages and phone numbers that could be scrolled through on the screen.

Business quickly realised there was a wider opportunity not just to advertise in Yellow Pages but additionally offer a range of business services. As a result the Minitel is now regularly used to buy plane and train tickets, make reservations at the theatre, opera and restaurants and in simple detail access product/service information on items ranging from insurance, retail banking through to shop opening times and prices. What Minitel cannot cope with as it is so under-powered is any detailed or complex transaction.

There is no doubt that the Minitel has successfully introduced all French consumers to the world of electronic information-gathering and purchasing for basic services. Use is widespread and knows no age or social–economic barriers. In fact technophobia has been replaced by a form of technophilia. The French generally love their Minitel and don't want to see it replaced. French businesses, too, are keen to preserve the Minitel operation as it keeps French consumers focused on French providers and keeps them away from the global competition found on the Internet.

Indeed the French government initially reacted with hostility to the Internet seeing it as another form of Anglo-Saxon invasion into French culture and the French way of life. In 1995 France's Centre for European Studies received wide publicity for its claims that 'French values are under threat . . . we cannot build Europe around the language of America'. At the same time France's Ministry of Culture hired staff to set up a special department to create and promote web sites in French.

By 1998, however, the French business and government communities were being forced to acknowledge the sweeping potential of the Internet and that if they did not invest and get involved they were in danger of losing out in a significant way in the race to the global electronic market. French Prime Minister Lionel Jospin has now called for a 'mobilisation of the French around the Internet'. Government is now starting to invest heavily in what Jospin describes as the 'information society of the future'. It's a tacit acknowledgement that the Minitel, such a successful pioneer, has had its day and it's now time for the nation to get rewired with more modern connections. As part of this drive, Jospin has 'ordered' every school in France to develop an 'Internet access plan' and is setting aside funds to create an Internet site at every one of France's post offices. There is a determination to adopt the Internet as a standard for all public information and a desire to encourage business to embrace the medium in the same fulsome way.

Indeed there need be no fears about an English-language invasion. There are a wave of new companies such as Globalink (ironically a US organisation) who are developing software that translates English e-mail and web sites into a chosen local language. While French business groups are expressing strong interest in this there will also be applications into German, Japanese and most other widely spoken languages.

*Germany*

Even without a Minitel system, Germany has still developed as one of the most advanced home-shopping countries in Europe. This is largely due to the presence of two of the world's largest mail-order companies – Quelle and Otto Versand – being based in Germany. Both these organisations have built substantial catalog, phone, fax, mail operations which are highly developed and popular. It's created yet another base and platform on which to more easily build a more sophisticated electronic shopping environment. CD Roms are replacing bulky paper catalogs, TV interactive shopping is developing well and Germany is one of the leading European countries wiring up to the Internet.

Clothing is the staple home order product, accounting for more than 50% of sales. Quelle and Otto Versand have learnt to target different consumer segments with tailored offerings that appeal to their needs. Like Next in the UK, they realise that mass mailing and communication has much less impact than getting into the real underlying detail of what will work with individual consumer groups and responding accordingly.

German home shoppers say their key reasons for buying in that way are (1) convenience, (2) the opportunity to choose at leisure, (3) shopping is stress-free, (4) comparing prices and products is easy and (5) they can get a customer-oriented individual service.

Home shopping has received a further boost of interest with the arrival of Quelle's own TV station called HOT (Home Order Television) which was launched in 1996 transmitting 24 hours a day. It has found a great freedom to broadcast as the German TV authorities have concluded that HOT is not television in the formal sense requiring regulation: 'It is a merchandising procedure and therefore subject to different laws.'

HOT is proving successful. It reaches into nearly 10 million German households and is encouraging further the German addiction to TV. Planning restrictions have long controlled German retail-store opening hours and forecasters predict the convenience of 24-hour, 7-days-a-week shopping will have particular appeal. Estimates are that at least 10% of all German retail turnover will be electronic by early in the next decade.

*Singapore*

Singapore is a compact island with a population of only 3.6 million. Computer penetration is high and the *World Competitiveness* report places Singapore among the top nations in the world in terms of strategic exploitation of IT, computer literacy and telecommunications infrastructure.

Since extensive privatisation of the telecoms network in Singapore, the industry is now well developed. The IT2000 master plan, set up by the National Computer Board (NCB) in 1991, seeks to develop Singapore into an 'intelligent island'. An integrated and extensively connected National Information Network (NIN) is being established to link computers and other information appliances in homes, offices, schools and factories.

PC ownership has risen from 27% of households in 1994 to 41% in 1998. Approximately 50% of these households are connected to the Internet. Public access to the Internet first became available in 1994, and there are now three Internet service providers to home users, namely Singnet (the first provider, launched in July 1994), Pacific and Cyberway (Table 4.13).

The Singapore government is looking far ahead in its commitment to the latest network technology. Singapore ONE (One Network for Everyone) is an initiative launched in July 1997 which combines two distinct but integrated levels. The first is a broadband infrastructure of high-capacity networks and switches which will break the current bandwidth barriers of current Internet technologies. Using the latest technologies, the Singapore ONE Network will easily handle high-resolution video, sound and 3D graphics, presenting businesses with the potential to develop fast services for customers. Connections will upgrade from the current standard of 33.3Kbps to 5–10 Mbps.

Construction began in 1996, and the network is scheduled to reach all areas of Singapore by the end of 1998. The government aims to have every home in Singapore connected to Singapore ONE by the year 2000, with around 20,000 households currently being cabled each month. At present, the programme is running as a pilot scheme (due to end in 1998), with over 100 applications developed and around 5000 homes and businesses actively using the network.

Table 4.13  Singapore Internet Growth, 1995–7

| Year | No. of Internet dial-up accounts | Growth |
| --- | --- | --- |
| 1995 | 25,000 | |
| 1996 | 129,000 | 516% |
| 1997 | 230,000 | 178% |

'The implementation of Singapore ONE makes Singapore the leader in the Asia-Pacific Region in establishing a nation-wide broadband network. In the next five years we will focus on establishing network linkages between Singapore and the rest of the world.'
(*Teo Chee Hean, National IT Conference and Minister for Education*)

\* \* \*

**When will it all reach critical mass?**

From Germany to Japan, Mexico to Malaysia business and consumers are getting wired into the electronic economy. Where there was initial reluctance as in France it has been replaced by a concern not to get left behind. Where there was hesitation as with grocery shopping (would consumers buy fresh fruit and vegetables?) there is now a growing recognition that the search for convenience, time-saving, leisurely, hassle-free shopping has strong appeal. Initial perceptions that this is just another distribution channel, like putting products into vending machines, are gradually changing. There appears little doubt that the appeal of electronic shopping will be far more widespread and will gradually become mass and mainstream.

When will it impact and take meaningful effect? In the research in this book the date of around 2005 has been established and this section examines that prediction in more detail. Certainly no company can now afford to ignore these developments and while some may argue the impact will be delayed in their particular industry sector there will be few arenas that ultimately will not be significantly affected.

Ignoring predictions can be a dangerous pursuit. This is especially so with the wealth of activity and investment in electronic commerce that is taking place. As history records and as Professor William Keep has pointed out, there have been some famous gaffes as various technological innovations in the past have been dismissed, their potential doubted, only for them to go on and fundamentally change society:

- 'The phonograph is of no commercial value.'
(*Thomas Edison, 1880*)

- 'Everything that can be invented has been invented.'
(*Charles Duel, Director US Patent Office 1899*)

- 'There is no likelihood that man can ever tap the power of the atom.'
  (*Robert Milliken, Nobel Prize winner in Physics, 1925*)

- 'Who the hell wants to hear actors talk?'
  (*Harvey Warner, 1927*)

- 'I think there is a world market for about five computers.'
  (*Thomas J Watson, Chairman, IBM, 1943*)

- 'There is no reason for any individual to have a computer in their home.'
  (*Ken Olson, President of Digital Equipment Corp., 1977*)

Electronic technology has been a major driver of change throughout the twentieth century and it's achieving impact in the market place at an ever quicker pace as supply-side competition rushes to translate ideas from the lab into commercial reality at affordable prices. Table 4.14 shows the increasing speed with which technology changes lives.

Beside technology, there are an extraordinary number of other factors fuelling the growth in electronic commerce generally and shopping in particular. Consumer time poverty, the general shift from a physical to digital world based on knowledge rather than assets, the convergence of companies from different sectors as multimedia develops, unusual levels of government intervention and support from the USA to France encouraging investment and providing central funds and resources, a rapidly developing infrastructure, with the USA again especially leading on PCs in the home plus Internet connections and advanced optical fibre cable links, and finally a new generation of people growing up computer

Table 4.14  Time for Technology to Penetrate the Mass Market

| Interface | No. of years |
| --- | --- |
| Telephone | 38 |
| Cable TV | 25 |
| Fax machine | 22 |
| Microwave | 13 |
| VCR | 11 |
| Cellular phone | 9 |
| PC | 7 |
| CD Rom | 6 |

*Source*:  Info Tech, Pac Tel Cellular, Coopers & Lybrand.

## Future Growth of Electronic Shopping 65

and 'Net literate' for whom using the technology is as commonplace and institutionalised as the telephone has already become. As a VP of Burton Group, the UK clothing retailer (already quoted), has recently commented: 'too many people are investing too much for it not to happen.'

With this bandwagon rolling, with this increasing momentum for change and advancement in commerce and life-styles it is no wonder forecasters in the know are predicting significant impact in just a few short years. A number of forecasts have already been mentioned but we can remind ourselves of perhaps some of the most compelling:

- 'By the year 2000 a third of the most attractive households will be ready to buy on-line.'

  (*McKinsey report*)

- 'Once the hardware and telecommunications infrastructure is in place an enormous range of services such as location independent shopping ... will likely be widespread by 2005.'

  (*Jagdish Sheth, Professor of Marketing, Gouizeta Business School, Emory University*)

- 'Electronic Retailing will grow inexorably through the next decade ... accounting for as much as 25% of US retail sales by 2010.'

  (*Professors Louis Stern and Barton A. Weitz*)

- 'On-line services will replace at least half the US citizen's average number of monthly shopping trips to grocery and related outlets by 2007.'

  (*The Consumer Direct Consortium of 31 US consumer goods companies*)

In addition to the research conducted in this book, some 38 different forecasts and predictions on electronic commerce have been reviewed and assessed to understand how other market commentators see the future evolving. These other forecasts come from a range of different individuals and organisations including practitioners and observers. The mean date that emerges from this combination of research is around 2005 (Figure 4.3). By around the middle of the next decade a majority of forecasters see electronic selling and buying reaching critical mass where a significant part of the population regularly shop electronically. The inputs come from information technology research companies like Gartner Group, International Data, general trend-spotting firms like Forrester and Jupiter, consumer market research organisations such as the Henley

**Figure 4.3** Predicting when ES will Reach Critical Mass*

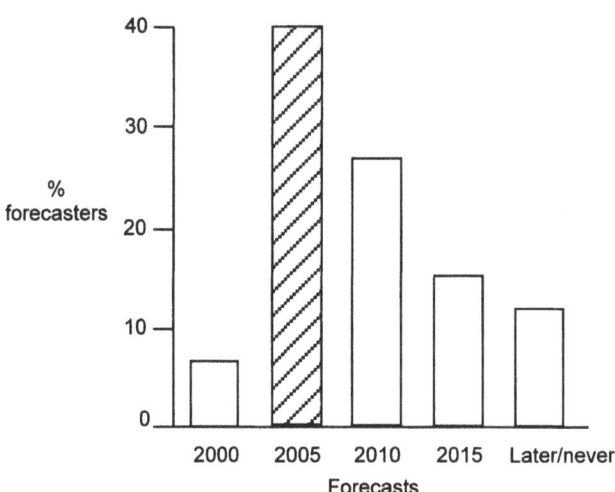

Note: *Defined as c.15% plus of consumers regularly shopping electronically.

Centre and AC Nielsen, consultancy firms like Ernst & Young, CSC and Andersen, companies such as IBM, Cisco, Karstadt and Chrysler and individual commentators like Professor Fred Phillips, Professor Jagdish Sheth and Dr David Diamond.

Of course no one can be absolutely certain as to when electronic shopping will really take off and achieve mass volume penetration. It does seem unlikely that retail sales over the Internet will have reached substantial levels by as early as the year 2000. But with technology developing at an exponential rate there is confidence that the next decade will begin to see the new electronic world taking shape, with business and consumers becoming increasingly familiar and comfortable operating in that way.

This development can be mapped out into five stages of evolution and maturity (Figure 4.4). It naturally follows the 'S'-curve pattern of investment leading to a wave of growth. That is typically followed by a period of consolidation and then further investment fuelling the next rise in growth and acceptance:

**Stage 1:** 'Initial hype': this was the period of early excitement and interest. It saw universal media comment and established high levels of awareness across the mass consumer base. But access was limited and so usage still low.

## Future Growth of Electronic Shopping

**Figure 4.4** Growth in Electronic Shopping: Key Stages of Growth, 1993–2007

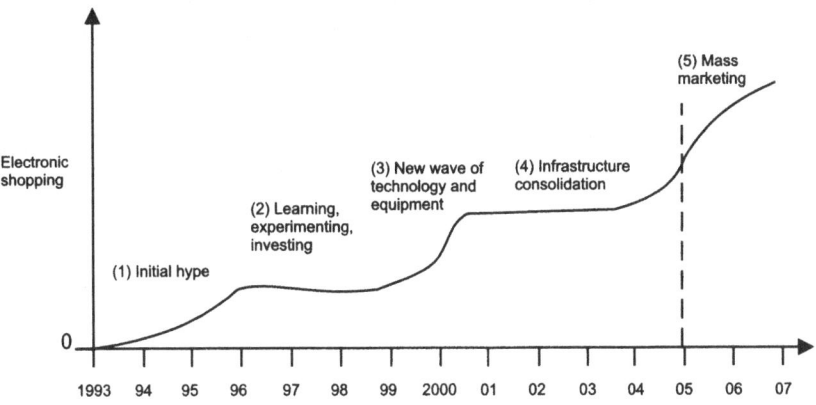

**Stage 2:** 'Learning, experimenting, investing': after the hype a gradual settling-in period has taken place where companies are learning how to make electronic shopping work for their consumers, establishing and finding the essential investments and setting up experiments and trials.

**Stage 3:** 'New wave of technology and equipment': the supply side is beginning to get its act together, the Internet begins to reach into the home through improved Web TV facilities, digital TV broadcasting scheduled for 1998–9 allows a range of new home-shopping channels to broadcast, more user-friendly equipment becomes available, mobile/hand-held technology develops bringing with it Smart Phones and PC Phones, PCs become cheaper and prices reduce rapidly following the pattern of other electrical equipment for the home. But it's all still not at mass consumer acceptance levels!

**Stage 4:** 'Infrastructure consolidation': companies can begin to meaningfully define the size of the opportunity – consumer expectations and demand are growing now the infrastructure is rapidly improving. The 1990s generation of Net kids is growing up and starting to shop more widely setting up their own homes. Efficient and widespread home delivery operations are built up and fine-tuned, equipment is designed for simple home usage (a typical equipment breakthrough being the remote control for using the TV), cable networks are extended, TV and satellite communication links enhanced.

**Stage 5:** 'Mass marketing': equipment is low-price and user-friendly, access is via the PC or TV or phone but is now sold as part of the package with every new TV, marketing is widespread, promotion levels are geared up for mass consumer appeal, it becomes the next piece of electrical gadgetry to have in every home, technology has learnt how to get speed and quality of transmission to acceptable levels through existing telephone wires, other communication links through satellite or the electricity grid are also now coming on stream, home delivery operators are well-recognised and developing a role as part of the local community, people are getting used to shopping where they want, how they want, when they want and they like the extended choice.

As Internet and other forms of electronic shopping reach wider levels of consumer acceptance so the user profile is changing. It's no longer the obsession of a few computer buffs, it's graduating beyond the hobby of the IT/PC-literate individual. By reaching into the home it reaches the family. It's not just dad upstairs in his study or den but mother and kids too.

It becomes part of the fabric of living, woven into life-styles, accepted and adopted as an essential tool for information-gathering, communication, obtaining services and helping with the shopping. It's the global version of a more sophisticated Minitel. It doesn't turn us into a nation of couch potato zombies, it doesn't leave us marooned at home, it doesn't destroy social activity, it doesn't stop people going out. On the contrary it gives us more time to be more sociable. It simply removes hassle and makes shopping more convenient. Why shouldn't that appeal to the mass consumer base?

What the 'S'-curve analysis in Figure 4.4 shows is that there will be clear periods of investment and consolidation leading to new waves of growth. By 2005, the supply side will probably be geared up, the infrastructure more in place, the fulfilment mechanisms better established, thus enabling the mass marketing of electronic shopping and true levels of mass consumer penetration to really begin.

By this stage we can be sure there will be a 'hard-core' user base who will be the regular and often dedicated electronic shoppers. They are the 15–20% of the population who in one research study after another – from A.C. Nielsen, Mintel, BMRB, the Henley Centre and many others – emerge as clearly wanting the convenience and time-saving. They have expressed a determined preference to take advantage of shopping

## Future Growth of Electronic Shopping

electronically as soon as it becomes available and accessible to them. They will be the first to get the equipment, log on, experiment and subscribe to the home delivery operation once it reaches their neighbourhood.

But they won't be alone. Following on their heels will be other groups of users, less dedicated but still looking to take advantage of the services on a more elective basis. By 2005 we might expect the pattern of electronic shopping users shown in Table 4.15.

Table 4.15 Possible ES Shopping Profiles

| Consumer acceptance (%) | % items shopped electronically |
|---|---|
| 1. 15–20% 'hard core' | 50–75% |
| 2. 15–20% occasionals | 25–35% |
| 3. 15–20% experimenting | 10–20% |
| 4. All the rest still irregulars! | 5% |

As this pattern of usage develops then some 15–20% of all retail shopping will be taking place electronically by around the middle of the next decade. Consumer shopping habits will be changing. Many will see the old days of pushing heavy shopping trolleys around as totally outdated. A new generation will be growing up and reaching maturity with different expectations. They will be familiar with the technology and want what's new and exciting. They'll be looking to forge their own lifestyles and patterns of behaviour and won't have time for quaint notions of visiting the grocery store to lug home the soap powder.

The challenge is truly set for all retailers and manufacturers. Given their positions in the market place they will be able to see this 'S' curve unfolding. Some for sure will stand back and wait and see, perhaps leaving it too late to capture the emerging consumer trend. Others will be playing a leading role, defining and shaping the future of their industry sectors. In the next few chapters we can examine the opportunities and strategies available to retailers specifically and later on we can also see how manufacturer suppliers can best respond to this new environment. Whatever the approach, the writing is on the screen.

\* \* \*

For a contrasting perspective, let's stand back for a moment from electronic shopping activity and consider the development of a different home innovation – the microwave. Here's an extract from the

*International Herald Tribune* telling the story of how the microwave took time to catch on and be accepted but eventually became an essential feature of every home.

> 'Eleanor Vavricek remembers the day she came home from work and laid her eyes on the brand new, boxy-looking contraption on her kitchen counter.'
>
> 'I thought, what is that, what in the world do I do with it?' Cook that's what, like never before.
>
> The new fangled gift from her husband was the new Amana Radarange microwave oven which made its debut in American kitchens in September 1967.
>
> 'It made my life a lot easier and simpler', said Vavricek. Many now feel that way too. Americans rank the microwave oven the no. 1 technology that makes their lives better. That puts the microwave ahead of the telephone answer machine and automatic teller machine which rank second and third.
>
> Yet Amana officials were nervous about introducing the microwave technology in 1967. Even in an age of astronauts and TV dinners they did not really believe the product stood a chance. They spent a long time educating wholesalers and retailers on how it worked before finally launching it. The company even hired home economists to make house calls to help first time buyers. Those home economists were on call 24 hours a day backed up by servicemen guaranteed to arrive within one hour in case of trouble.
>
> It took time to develop. Ten years after introduction in 1977 it still could only be found in around 1 in 10 households but after 1977 it began to take off to the nearly 90% penetration it has today.

# 5 The ES Test: How to Tell How Much Your Business is Going to be Affected

Which products and services are going to be most affected by the trends and developments in electronic shopping ('ES')? Are there some which are so obviously susceptible that they can be readily sold on-line without any noticeable loss of value-added? Can those be easily identified? In contrast are there other products whose characteristics demand such a personal involvement and physical interaction that electronic selling is never likely to make much impact, except among a dedicated few?

Where product characteristics do point in the ES direction then must existing manufacturers and retailers give up on established traditional distribution routes and move quickly into the virtual world? Can they buck the trend? What are their various options and how can retailers and manufacturers begin to make sense of these market developments? Can a framework be devised that will help pin down the extent of the electronic threat, and opportunity, and the likely strategic responses required?

The 'ES Test', described in this chapter, provides a simple three step approach to help address these challenges. It applies to all types of products and services in all industry sectors. It can be used by retailers, by manufacturer suppliers, by any company participating or looking to participate in some part of the retail distribution chain from factory gate into the consumers' home. As the test is described, 'product' will be used to refer to both product and service industries.

The three steps in the test (see Figure 5.1) are:

1. **Product Characteristics**
2. **Familiarity and Confidence**
3. **Consumer Attributes.**

- 'Product Characteristics' is about the product's primal appeal to the senses. Does it need to be physically touched or tried before buying?
- 'Familiarity and Confidence' looks at the degree the consumer recognises and trusts the product, has tried it before and is confident in repurchasing it.

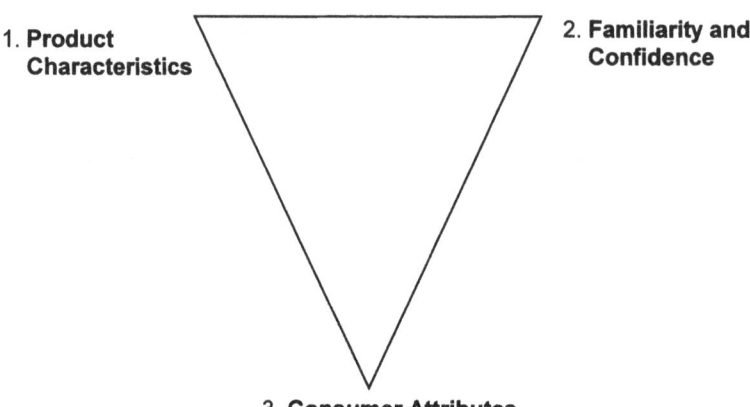

**Figure 5.1** The 'ES Test'

1. **Product Characteristics**
2. **Familiarity and Confidence**
3. **Consumer Attributes**

- 'Consumer Attributes' considers the consumer's underlying motivations and attitudes towards shopping. Will they carry on visiting the shops even if a product naturally lends itself to electronic purchasing or will they be among the first to take advantage of the new medium?

Before describing and showing how best to use each part of the test it's worth emphasising that all three steps must be taken and checked out. A full evaluation needs to be made. Put simply, a product may seem to have obvious 'electronic characteristics' and be a familiar item in most consumers' shopping baskets. However some of those consumers may themselves just not be ready yet for electronic shopping, be reluctant to deal with new media and may respond in only a very limited way to on-line selling.

For example, a 35-year-old professional may be perfectly confident in buying insurance over the phone but a 65-year-old retired blue-collar worker may likely have a very different attitude and prefer face-to-face personal advice. Equally a busy mother with young children may relish the convenience of the family's weekly groceries being delivered to her home but a young single adult newly arrived in the city might look forward to the social interaction of a visit to the grocery store.

While superficially straightforward these examples simply demonstrate the importance of not rushing to conclusions based just on the product's characteristics. Understanding the target consumers and their likely responses and motivations will be as important and may well drive final decision-making. Not surprisingly success comes best from being market and customer-driven, rather than just being product-led!

## 1. Product Characteristics

Every product has an innate set of characteristics that makes a primary appeal to the consumer's senses. This appeal is traditionally ascribed to the five senses of sight, sound, touch, taste and smell. Some products appeal primarily to one of these senses such as a painting, music, perfume. Others, such as buying food, clothing, a car or a home will call up a whole mix of senses and appreciation

Each product's principal senses can be identified and at the same time categorised as lending themselves to electronic purchasing – or not (Table 5.1).

Products which appeal to the senses of touch, taste or smell are *prima face* not suitable for electronic selling. Consumers expect to be physically involved in the purchase decision and want to handle, try and experience it before deciding. This is especially true if the product is new and has not been bought before. Other products that appeal mostly to sight and sound more naturally lend themselves to being bought over the Internet or other electronic connection. Indeed as the quality of Internet communication improves so sight-sound-based products can be expected to see increasing share of sales on-line. The product's value-added can be effectively communicated in a 'virtual environment' and the consumer loses little through lack of physical contact or involvement.

Looking at the early consumer surveys, some of the best sellers over the Internet have been basically sight/sound products that most naturally and immediately fitted into the new medium. Books, computer hardware and software, music CDs, basic financial services, information services generally all have a distinctly sight/sound appeal and suppliers have found it relatively straightforward to communicate their basic product proposition (Figure 5.2).

**Table 5.1** Five Senses: Primal Product Appeal

| Sample Products | Sight | Sound | Smell | Taste | Touch |
|---|---|---|---|---|---|
| Music |  | X |  |  |  |
| Perfume |  |  | X |  |  |
| Books | X |  |  |  |  |
| Computers | X |  |  |  |  |
| Autos | X | X |  |  |  |
| Food | X |  | X | X | X |

Higher ES potential ⇐ ⇒ Lower ES potential

74                    *e-shock*

**Figure 5.2**   Physical vs. Virtual

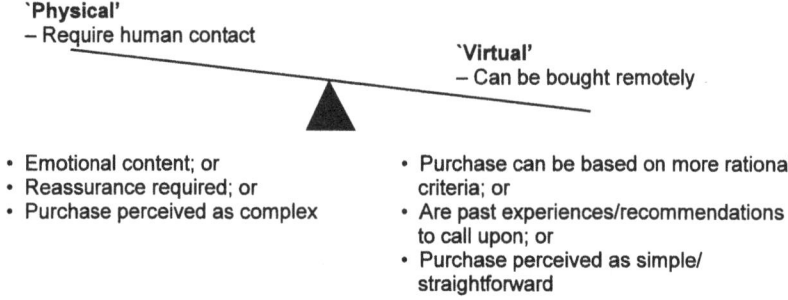

- Emotional content; or
- Reassurance required; or
- Purchase perceived as complex

- Purchase can be based on more rational criteria; or
- Are past experiences/recommendations to call upon; or
- Purchase perceived as simple/ straightforward

We can go further and map out (Figure 5.3) the whole list of retail consumer products defined in Chapter 2 and immediately establish an initial view of which products have a principal appeal to selling and buying electronically.

As the map in Figure 5.3 shows there are fewer touch/taste/smell-only products. The majority are sight/sound in nature and it would seem that a large part of the retail scene will likely have a basic responsiveness to electronic shopping activity.

**Figure 5.3**   Product Characteristics: Responsiveness to ES

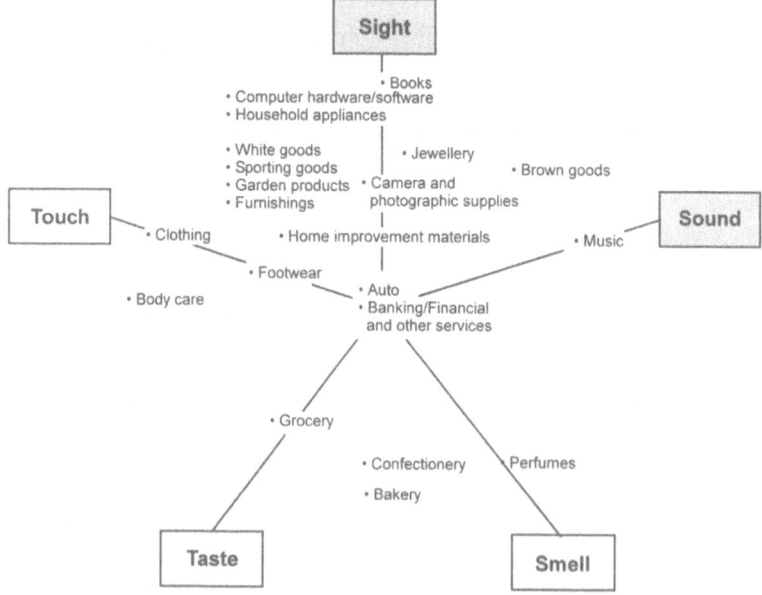

But, before conclusions can be drawn and before companies operating in any one product area rush to build their on-line franchises, it is necessary to drill down into more detail. At an individual product line level we can look for example within Grocery food (Table 5.2) and show that while some products certainly do share the generic Grocery 'Touch, Taste, Smell' appeal, others don't. Some are so packaged up that touch taste smell sensations are lost.

Table 5.2  Product Appeal for Individual Grocery Lines

| Grocery products | Sound | Sight | Touch | Taste | Smell |
|---|---|---|---|---|---|
| Strawberries | | | X | X | X |
| Bananas | | | X | | |
| Chicken | | X | X | | |
| Breakfast cereals | | X | | | |
| Cookies | | X | | | |

Where a grocery or other product is packaged it does limit the sensual opportunity. The only criteria for purchase are 'like the look of the packet', 'have tried other products from same company', 'like that combination of ingredients'. Criteria such as these can be evaluated as easily on-line or via a home catalogue/CD Rom as they can be in the store.

The same analysis can be done for clothing. Basic items like tights, undergarments, socks, handkerchiefs and shirts are often sold in boxes or wrapped up. The manufacturer and retailer are trying to make the purchase easier but in doing so they have removed much of the emotional content in making a purchase decision and made physical interaction before buying unnecessary – and with all the packaging nigh on impossible. Marks & Spencer, one of the world's leading clothing retailers, packages up many of these basic items in its stores to encourage easier purchasing. It is noteworthy that it is already experiencing demand for clothing items to be included in its home delivery services.

As suppliers and distributors reflect on the challenges of the electronic age one area they will need to re-examine is the way in which products are presented. If they still want consumers to visit the store the emotional and physical experience needs to be enhanced and reinforced at every turn. Things that get in the way like plastic wrapping and boxes may need to be removed and the product itself reinvested with a sense of fun, of being interesting and different, for example, in the way it is merchandised and the other products it is presented and associated with.

Looking again at the Product Map in Figure 5.3, one other important observation is how services like Banking and other financial transactions don't naturally belong to any one of the five senses. It's as though their key characteristic lies with a broader sixth sense which might be described as the 'intellect'. Buying these products is typically a rational analytic process, not relying solely on sight or sound. It rather draws on a whole range of derived experiences and inputs which are then assessed in the mind and then a buying decision is made.

The more the product or service appeals to this broader 'intellectual' assessment, in contrast to something more physical or emotional, the greater its electronic selling potential. We could even redraw the product characteristics map (Figure 5.4), adding this sixth or derived sense as its centre. Where a product has a particular primary sensory appeal it can be plotted firmly along its axis. Where the product or service has this broader 'intellect' characteristic it is drawing on the whole range of sensory experiences and can be mapped across the centre.

The map can be redrawn in this way to help pinpoint as clearly as possible what are the product's basic characteristics and its electronic appeal. Any conclusions at this stage still need to be kept on hold, however, while the next two steps of the test are checked out.

**Figure 5.4** Product Characteristics: Adding a 'Sixth' Sense

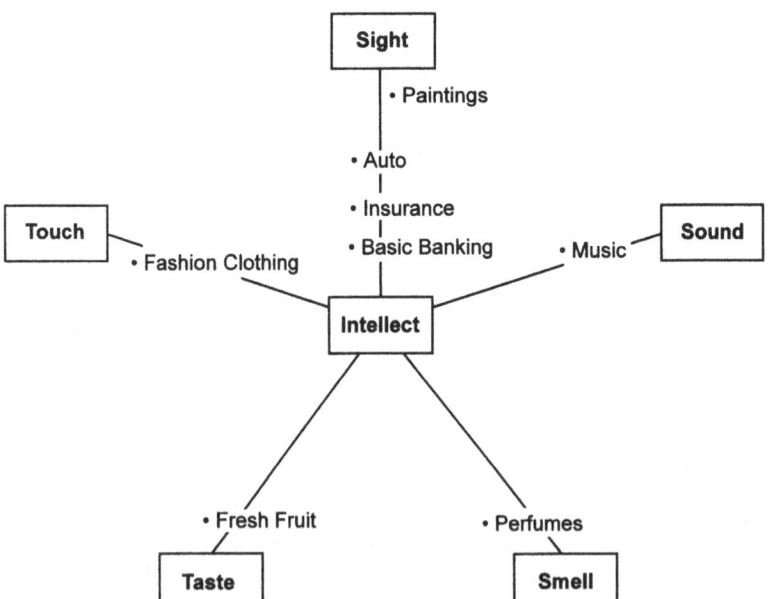

## 2. Familiarity and confidence

The second step in the ES Test looks at how familiar and confident customers are in purchasing an individual product. Have they used it before? Have they purchased it previously? Have they had a good experience with it and are pleased to buy it again? The greater the familiarity and comfort with a product the less the need to be physically involved in its repurchase and so the easier to buy electronically (Figure 5.5).

Familiarity and confidence can typically be found among all types of everyday purchase:

- Basic Foodstuffs – cereals, dairy products, potatoes, bananas, cookies
- Drinks – beers, colas, fruit juices, wine
- Basic Clothing – undergarments, socks, tights, regular shirts, and blouses, e.g. for the office, workwear.

These products are bought every week and consumed every day. An Indiana University research study showed that c.80% of a customer's average grocery shopping basket was made up by replenishment items. There is a high degree of familiarity with much of the weekly shop. This would suggest a wide range of what's currently sold in a supermarket for example, could equally as well be sold by catalogue/CD Rom or on-line and ordered from home without visiting a store.

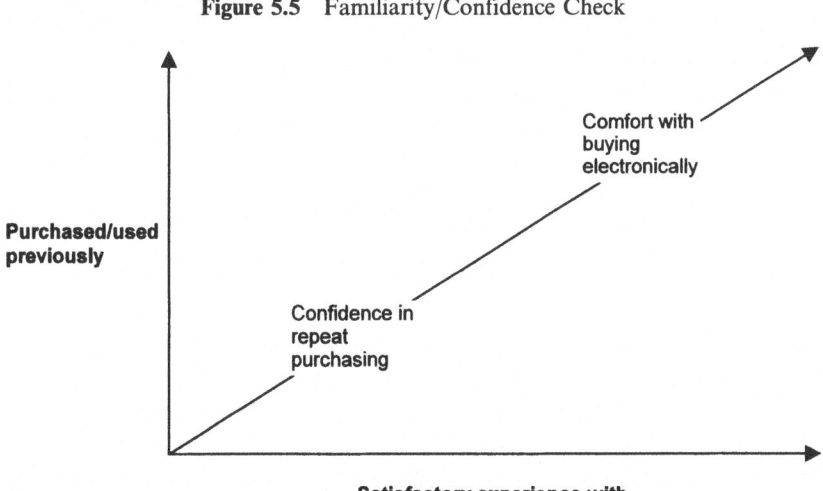

**Figure 5.5** Familiarity/Confidence Check

**Table 5.3** Top Ten Brands

| | | | |
|---|---|---|---|
| 1. | McDonald's | 6. | Gillette |
| 2. | Coca-Cola | 7. | Mercedes-Benz |
| 3. | Disney | 8. | Levi's |
| 4. | Kodak | 9. | Microsoft |
| 5. | Sony | 10. | Marlboro |

*Source*: Interbrand.

Consumer familiarity and confidence is especially high where there is a strong brand name on the product which has a reputation for reliability and trust. Look at today's list of strongest brands (Table 5.3) – would consumers find it difficult to buy these electronically? Would they still feel the need to physically interact with the product before buying it? Isn't the product's established reputation sufficient?

Among other things, a strong brand creates a depth and breadth of consumer franchise that cuts across ages, demographics and geographies. A strong brand builds commitment and loyalty that persuades consumers to pay more money for it and sometimes go out of their way to get it because of its perceived value and reliability. It's hard to see how virtual or electronic purchasing – once the infrastructure is in place – will present any barrier to consumers buying established branded products.

Those companies who have invested in their brands over time have put themselves in an excellent position to take advantage of future changes in shopping habits. As consumers do become more familiar with buying on-line and as the number of store visits reduce so there will be a premium on established reputations. For consumers learning to navigate their way through cyberspace the easiest purchases will be the products they recognise or the brand names that have the track record.

For new companies wishing to 'break in' there will be renewed emphasis on brand-building and brand-sustaining skills and investments. Even today the physical market place is crowded, saturated in some categories, with many products seen as commodity-like. In such an environment branding may be the only meaningful differentiator that makes one product stand out and become the first choice for consumers. Many companies have long struggled with the challenges of effective brand-building, concerned about how much it costs, struggling to see the pay back, often compromising with half-hearted attempts which just don't break the noise barrier or achieve any significant impact. To go further and establish a lasting reputation requires a profound and enduring commitment to brand-building which only a few companies in each sector seem prepared to make.

This branding challenge will become even more acute in an electronic shopping environment where physical interaction is reduced and where the product's qualities and benefits must be distilled and captured in a way that can be communicated over the wires. It will be more difficult because consumers will now have greater control over what virtual stores and aisles they visit. They are no longer forced to push the shopping cart up and down every aisle past every one of the 20,000 or more product lines on display. They will instead choose where to shop, what categories to explore and how much time they want to spend looking at what's available, comparing prices or getting more information on new product lines.

There is more on the challenge of building consumer confidence in Chapter 14 but there will be an undoubted correlation between strong, recognised brand values and success in electronic selling. Indeed, consumer familiarity and confidence with the product can be assessed now to see if it would stand up well in an electronic environment. Is branding an important purchasing criteria for products generally in a target category? Does the product have sufficient immediate recognition?

A set of questions can be laid out to enable those concerned with product brand strength/consumer familiarity and comfort to quickly assess it and start identifying what are the gaps and investments required to get the Brand to where it needs to be (Table 5.4). These questions can form a basic piece of consumer research and act as a familiarity/confidence check for the product in its market. The aim is to score each answer 0 to 10, with a strong Yes scoring 10 and a strong No 0. 10 would indicate the highest levels of brand strength, 0 the lowest, 5 an average. Scores less than 60 here are inadequate. The strong brand-builders mentioned in Table 5.3 would likely score 80 plus, setting a standard and benchmark for what is needed to succeed.

Building consumer familiarity and confidence is an important success hurdle in any market place, whether physical or virtual. It is likely to be even more important in cyberspace. If a product's characteristics demonstrate ES potential then there may be no choice about building consumer confidence into a recognised and lasting brand name. It may well be that branding has played only a limited role in the sector in the past but the new electronic age will force would-be sellers and distributors to move more forcibly down that path. If one company does not respond and seek to exploit this opportunity then for sure rivals will eventually see the gap and come into exploit it – in the same way that Amazon.com has muscled into the books market and immediately given some branding lessons to sleeping rivals.

**Table 5.4** Consumer Familiarity and Confidence: 'How Strong is the Brand?'

1. Do consumers, without prompting, name the Brand as the leader in its category/sector? _____
2. Is the lead vs the next-mentioned Brand name significant? _____
3. Have a high proportion of target consumers purchased the product at least once? _____
4. Is there a high proportion who would count as regular users? _____
5. Is usage of the product typically highly satisfactory? _____
6. Would target consumers say they trust the product? _____
7. Would target consumers agree there is no need to check the product out before purchasing it again? _____
8. Are intentions to repurchase at high levels? _____
9. Do consumers see a clear difference in value vs. other competing products? _____
10. Is the ongoing brand investment by the company higher than that put in by competition? _____

This Step 2 in the ES test sets out an important check on the product's position with the consumer. It applies to every product and service. It's not just relevant for basic everyday purchases. Whatever the item, if there's ES potential then building confidence can be the key enabler to successful electronic selling and distribution. Even fresh fruit, long thought to be an unlikely electronic product, can be sold effectively on-line (as Peapod and others have demonstrated) where the seller establishes a reputation of trust and reliability on the freshness and quality of products sold.

### 3. Consumer Attributes

If the product's characteristics have electronic appeal, if consumers are familiar and confident in buying it then its ES potential is clearly high. But would consumers *want* to make the purchase electronically? Would they still prefer to visit the store? What type of consumers is the manufacturer targeting? What's the profile of those who are visiting the store? Certainly many consumers will respond to the lures and advantages of new, more convenient ways of shopping. But it is equally sure that others will be uncomfortable with the new medium and simply prefer to stick to their old and traditional ways of doing things. While traditions may eventually die out it can take a long time and companies need to know enough about their consumer to identify their electronic market – how big it is now as well as how it might develop long-term.

**Figure 5.6** The Triple 'I' Model

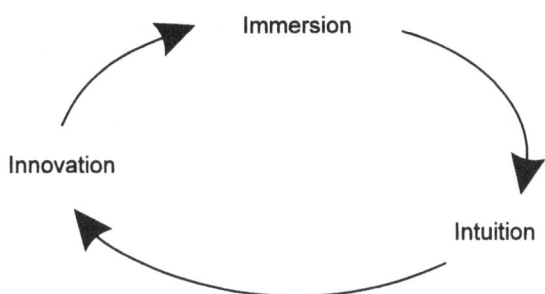

The key to building this understanding has already been referred to. It's through so deep-rooted a market immersion that a company really does understand who are its customers, what are their needs now, how they will develop in the future and the degree to which they'll respond to new purchasing options. This approach to appreciating consumer profiles and attributes can be captured in the simple 'Triple I' model in Figure 5.6.

Dig deep enough into the market place so that it's possible to intuitively understand the consumer and be totally in-tune with how they are thinking and behaving. Through that process it is easier to gain insights on how new initiatives will be received, what will work and what won't. With that appreciation new ideas will emerge with a stronger chance of succeeding and there will be stronger foundations on which to establish a reputation for leading innovations in the market.

This sort of approach has been adopted meticulously by leading consumer research organisations like A.C. Nielsen, Mintel and Verdict. A.C. Nielsen, for example, has a panel of c.10,000 consumers who are regularly reporting back to the company on tastes and preferences and giving input to what initiatives might work for them. We shall examine their research in detail in Chapter 9 but they identify a number of underlying trends when assessing the consumer base.

Consumers, for example, can be categorised into sub-groups. One such group are called the 'frenzied copers' who will be responsive to any initiative which can ease the pressure and make shopping easier and more convenient. Another group is the 'habitual die-hards' who are stuck in their ways and are reluctant to change. A third group is categorised as 'mercenaries'. They are searchers after value – either better price or better service – and they will respond to whatever proposition best offers that.

In sum, these categorisations show there are six consumer types (Figure 5.7). For each category it is relatively clear what their likely response to

**Figure 5.7** Consumer Categories: ES Responsiveness

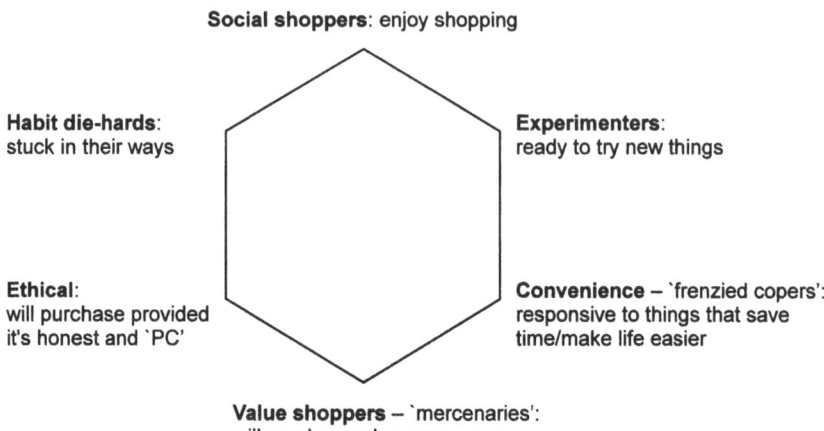

*Source*: A.C. Nielsen, Mintel, Henley Centre, Kalchas.

electronic shopping will be. Of course consumers aren't black and white and they don't fit in just one box so they can be neatly pigeon-holed by market researchers. The categories blur, consumers behave differently, change their attitudes, will be in one category for some products and services and in another category for others. But this research does identify the main underlying motivations and provides a framework to help determine the extent to which a company's current consumer base will likely become electronic shoppers:

- **Social shoppers** – these consumers enjoy visiting stores and take pleasure in shopping. It's an opportunity to get out of the house and meet people. Some stores have identified this trend as being particularly strong among their target consumers. For example, they have set up 'Singles' Nights' where the store stays open late and encourages shoppers to meet and socialise. Commentators have also observed how the bigger stores have become meeting places for people generally – they've become the community centre, often fulfilling the role previously occupied by the church or village hall.
- **Experimenters** – they are happy to try new things and are comfortable in doing so. They have been the first to use and purchase over the Internet. This category has a particularly strong representation among the growing Net generation of young adults who have grown up with

new electronic connections and have already established their comfort levels in using them.
- **Convenience shoppers – 'frenzied copers'**, they suffer most from time poverty and find shopping generally a chore. They represent the stereotype of the busy parent rushing to do the food shopping with reluctant children in tow on the way home from school who will grab at any initiative that can meaningfully improve their life-style, especially if there is little or no extra cost.
- **Habit-bound die-hards** – they will be among the last to change, preferring traditional means of shopping, reluctant to try things like computers. These will be the last to join the electronic revolution and will only come on board once the technology is so established that it becomes totally simple and accessible.
- **Value shoppers – 'mercenaries'** – the search for value is all pervasive but these place it at a premium. They will respond to any market place initiative which appears to offer the best combination of product quality, price and service. If the same product can be more easily obtained electronically at the same or lower price with similar or better levels of service back up, then they will be quick adopters of electronic shopping facilities.
- **Ethical shoppers** – there may be a minority who put ethics as their number one criteria over and above these other shopping motivations but this group cannot be overlooked. Their decision-making is likely to be indifferent to whether a product is sold physically or virtually, provided it is done in an honest and 'PC' manner.

Manufacturers and retailers can use this categorisation to evaluate their own consumer base. How many of their customers fit into each category? Is there much change between the categories and can their consumers be properly classified in this way? This type of evaluation is the start of the detailed consumer segmentation that's needed. Only by understanding their consumers in this detail can a company decide what type of electronic response is required.

The product characteristics may show ES potential, the familiarity/confidence check might get high scores but if the target consumers are all 'habit-bound die-hards' and 'social shoppers' there may be little point in rushing down the ES path. Equally product characteristics may show limited ES potential and familiarity may be low but if target consumers are experimenters and 'convenience–frenzied copers' then there will likely be interest and demand for electronic buying facilities ready to be exploited.

**Figure 5.8** ES Segmentation

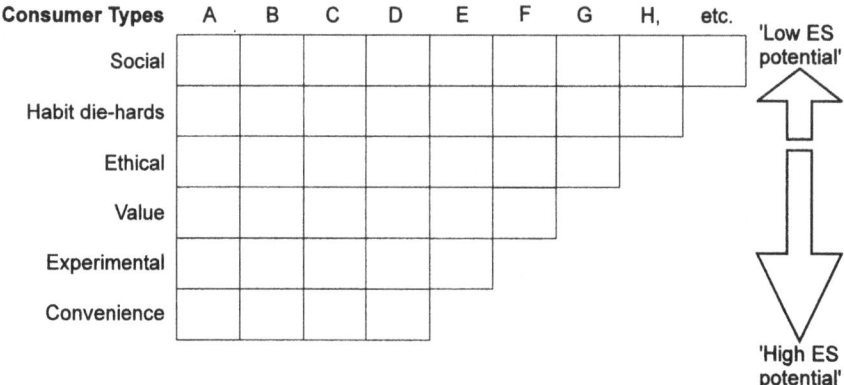

A company with a range of products should draw up a segmentation grid (Figure 5.8) to identify what types of consumers it has in each product category so it can begin to pin down the most appropriate ES response and action plan.

For each product category how many current consumers are in which consumer type? Where are the majority, who are the more frequent purchasers, which spend more or buy more of the product, which consumer types are under-exploited (e.g. are there many more 'frenzied copers' out there than the company currently attracts?) and with which group lies the greatest future potential?

These questions can be answered and the consumer base and its potential quantified. But no company can expect to succeed in the new electronic age without conducting the analysis. It can provide a clear view of the ES potential and determine whether and how best to respond to the electronic shopping revolution.

\*  \*  \*

### Summary and scoring

How should the three steps in the ES Test be brought together, the evaluation synthesised and the learning distilled? A simple scoring system can be introduced which adds up the total ES potential and consumer interest. A high score will evidently point the way for urgent appraisal of the ES opportunity and action to capture it. A low score suggests the ES potential is not yet fully formed and a different strategy will be required. The next two chapters will discuss strategic options in detail dealing with the retailer situation first. Chapters 12 and 13 will examine the position from the manufacturer supplier's standpoint. All the various strategic options require the key inputs of the ES Test.

Each of the steps in the test will contribute a score toward the total and can now be considered in turn.

## 1. Product Characteristics

The more physical the product's primal appeal, the lower its ES score. The more virtual its appeal, the greater potential for electronic selling, the higher its score. Scores out of 10 (Figure 5.9).

**Figure 5.9** Product Characteristics' Scoring

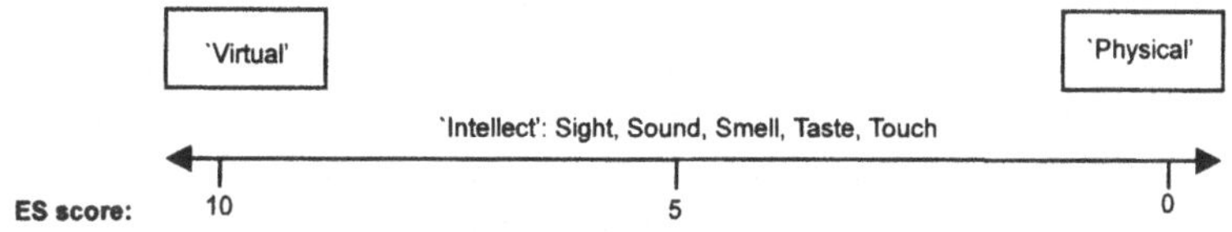

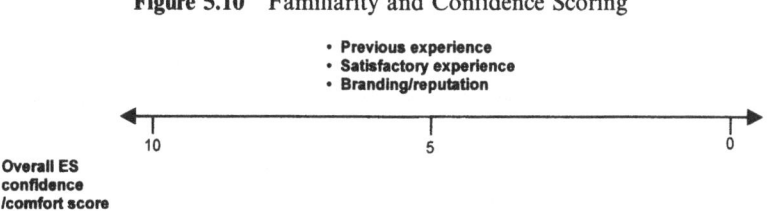

## 2. Familiarity and confidence

A research check on familiarity/confidence levels was set out in Table 5.4. The research boils down to three key conclusions around previous usage, satisfactory usage and general familiarity with the company or product's reputation (branding) (Figure 5.10).

**Figure 5.10** Familiarity and Confidence Scoring

- Previous experience
- Satisfactory experience
- Branding/reputation

```
        10            5            0
Overall ES
confidence
/comfort score
```

## 3. Consumer Attributes

As the three steps have been discussed it has become clear that the consumer's attitude toward electronic purchasing can be the deciding factor. Their interest and readiness in shopping electronically can override the conclusions from the other two steps and demand a response that meets their specific need. As a result the scoring for consumer attributes needs to be weighted so it adequately reflects the importance of the consumer driver. The suggested scoring in this part of the test is out of 30. The maximum score available then for the whole test is 50.

The key scoring measure for Consumer Attributes is the extent to which the target consumer base for the product or service is responsive to electronic shopping. How many dedicated convenience shoppers vs how many habit-bound die-hards or social shoppers? What's the overall proportion and is it of a sufficient extent that it provides a meaningful market place and level of consumer interest for selling electronically (Figure 5.11)?

**Figure 5.11** Consumer Attributes' Scoring

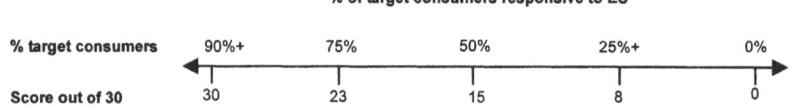

To illustrate this scoring for different product categories, we can take a generic approach based on average market place and consumer data and provide an initial suggestion (Table 5.5) as to which categories will likely have the highest ES potential.

Any score that's significant, especially if it's over 20, is demonstrating significant future ES interest and potential. Consumer attribute scores by themselves are unlikely to be high right now given that the general development of electronic shopping is still in its early days and many consumers have yet to gain easy access to the new medium. As that consumer access grows and widens so a higher proportion will develop their electronic interests and 'attribute' scores will start rising dramatically.

Table 5.5  Products with Likely High ES Potential-Scoring

| Sample products | 1. Product characteristics (0–10) | 2. Familiarity and confidence (0–10) | 3. Consumer attributes (0–30) | Total out of 50 |
|---|---|---|---|---|
| Basic grocery e.g. branded cereals and cookies | 4 | 8 | 15 | 27 |
| Basic household e.g. branded cleaners and powders | 8 | 8 | 15 | 31 |
| Basic clothing e.g. socks, tights, undergarments | 4 | 7 | 8 | 19 |
| Drinks e.g. branded beers, soft drinks | 4 | 8 | 15 | 27 |
| Financial Services e.g. auto insurance, standing orders, mortgages | 10<br>10<br>10 | 5<br>6<br>1 | 8<br>15<br>4 | 23<br>31<br>15 |
| Travel e.g. airline tickets, | 10 | 6 | 15 | 31 |
| Hotels | 7 | 6 | 8 | 21 |
| Books | 8 | 7 | 23 | 38 |

\*   \*   \*

This appraisal and scoring should be carried out on a regular basis. The market place is dynamic, things are changing rapidly and a low overall score can very quickly increase. Those who watch their market places carefully and evaluate the trends rigorously can put themselves in the best possible position to respond electronically if and when demand grows. They can ready themselves now, lay their plans, set their strategies, prepare their investments. Doing nothing, unless the score is close to zero, leaves a company vulnerable to fast-moving dedicated competitors or new entrants. Laying the groundwork will build the platform on which to succeed in the new electronic age.

# 6 How Can Retailers Respond?

The ES test will demonstrate the potential for a product or service to be sold on-line. It will help identify likely levels of demand and market size. It can be repeated regularly to watch for trends and other changes in people's readiness to buy electronically. It establishes a framework for determining right now whether and how best to respond.

If the indicators show significant ES interest then what are the response options? What can companies do about it? Must they embrace the electronic world completely, move out of existing retail distribution and deal direct and individually with each end-user? Are they tied into this strategic response? Or can they buck the trend instead and revitalise their proposition so that consumers will be content to carry on with their present shopping patterns? Are there any 'midway' strategies which do respond to electronic demand but which still aim to satisfy and serve traditional shoppers?

These next two chapters and the appendices focus specifically on the retailer dilemma and the strategies and opportunities that can be pursued. Chapters 12 and 13 look at the manufacturer supplier situation and examine their particular options. For both groups of operators there are a range of ten strategies that can be identified, each of which can provide a way through the current challenges and set a path for successful business-building into the next decade. But it's worth emphasising that the key, whichever strategy is chosen, will be its rigorous, persistent, fully resourced and funded implementation. In this highly competitive, fast-changing world, ambivalent half-hearted responses will always be in danger of being quickly overtaken by more dedicated rivals.

Retailers of any substance have already taken the first few steps in learning how to deal with this new era. They have been busy trialling and experimenting and in a low-key way are comfortable with making occasional mistakes. UK retailers like Argos, Sainsbury and Barclays Bank all launched web sites with a great fanfare in 1995 only to find, through subsequent research, that the web sites were seen as badly laid out, access was limited, quality of communication poor and consumers quickly disappointed with what was available. For example Argos included only a few of their household product lines, Sainsbury sold only wine out of its 20,000 products and Barclays Bank promised it would sell things but initially gave no time or indication as to when.

But retailers are using this learning positively. Web sites have been relaunched, phone fax ordering speeded up with more efficient call centres and order-processing systems, home delivery improved and expanded and significant effort put into getting chilled or frozen products to arrive at the right temperature at the right time. After the initial hype and unwarranted expectations there has been a settling-down period. Retailers have established much better resourced and skilled project teams. They have realised this may be more than a small incremental distribution channel. They are going back to the drawing board.

How well placed are retailers to deal with these new challenges and what will the retail environment of 2005 and beyond look like? Do retailers have the basic skills and capabilities to develop a competitive response? Will they need to ally and partner with others to get there? And what's the likely shape of the retail market they should be aiming at in the future? Will there still be high streets and shopping centres as we know them today or will some environments emerge as more attractive than others? Can small or medium-sized high streets with only half the range and choice and experience opportunities of a large shopping mall survive?

This chapter looks at both these issues – retailers' ability to respond and the future face of shopping – and establishes some of the parameters guiding retailers' response options.

**Ability to respond**

Retailers should not be fighting shy of confronting and dealing with the new electronic world. Quite the contrary, for especially in the last fifty years they have demonstrated a remarkable capacity for innovation and leading edge thinking. In particular leading retailers like Marks & Spencer, Tesco and Wal-Mart have themselves shaped consumer buying patterns and behaviour and at their best have led fashion and demand. Many have a history of being market drivers.

The most far-reaching move this century was the development of self-service. It was first pioneered in the USA during the depression when vacant warehouses were converted into supermarkets selling groceries with the minimum of customer service. It was introduced into the UK by Sainsbury in 1950 (see plate 1) and immediately impressed customers with the experience of choosing for themselves at their own pace. 'It is irresistible and seductive' recorded the journal *Home News* which reported on this new venture. Even the then British Prime Minister Harold Macmillan was moved to comment: 'it's a very clean and most ingenious way of serving the public and doing business'.

And retailers have of course continued to build on improving the supply chain. The past fifteen years, especially since the early 1980s, has seen significant changes in many areas. Much of the effort has gone into improving internal efficiencies in systems, warehousing, outsourcing of transportation, enhanced stock and ordering measures, EDI (electronic data interchange) between retailer and supplier to speed up invoice and order processing, EPOS bar coding and ECR (Efficient Consumer Response) for lower costs and more efficient management of the various stages in the supply chain. All these have had significant impact in reducing costs, ensuring stock availability and providing more efficient and timely distribution of products to the store.

These supply chain efficiencies have also been translated into benefits for the consumer. Vast improvements in freshness and quality have been achieved, sell-by dating systems have provided reassurances and guarantees of product suitability and packaging is now not only brighter but better preserves the product. In keeping with these developments the more sophisticated retailers have aggressively introduced their own private label in an attempt to further improve quality and value. In the UK especially private label has been so successful that leading retailer Sainsbury has nearly half its sales under its own banner and Marks & Spencer has an extraordinary line-up of manufacturer suppliers providing it with product under the St Michael brand name.

By the mid-1990s retailer supply chain and product innovations were beginning to peak. Where else to innovate, how to continue to excite, pull in customers and keep up the sales momentum? With the USA and UK especially saturated in many urban areas with retail selling space, retailers are in a position where they have to keep adapting, improving and changing if they are to sustain their position in the market and continue to grow shareholder value.

The latest area for innovation is utilising the total sales space to provide complementary ranges of services. Retailers now are looking to encourage consumers to begin to see the major stores as a one-stop shop, somewhere to spend an extended period of time and carry out all the basic shopping requirements. Retailers are doing this sometimes by themselves and sometimes in partnership. They are building service offerings ranging from gas stations through to banking:

- Tesco now offers its customers everything from its basic supermarket grocery food items through to Calvin Klein underwear, Armani socks, popular CDs and children's toys and board games. Its ever-increasing product range has been one of the major drivers of its growth, picking market share from other retailers' territories. In 1997 Tesco announced the trial of a new format store called Extra. It is a hypermarket size with 105,000 sq. feet, almost double the size of its current average store. The floor plan is laid out by department and has been extended to make way for numerous new departments.
- Supermarket banks were a new phenomenon at the start of the 1990s but now they're visible everywhere. In California, for example, Wells Fargo has built a network of 700 branches and in-store kiosks, Bank of America has 200 supermarket branches. All the major food retailers have tied up with banks. Kroger is linked with Nations Bank, Publix with Sun Trust. Bank of America is tied into different food retailers in different states working with Lucky stores in California and Jewel/Osco in Illinois.

   Supermarket banking is not confined to the USA. In the UK it only really started developing through 1997 but again all the major food chains have linked up with UK domestic banks. In Germany, Commerzbank had 30 branches in different supermarkets by mid-1998 and Deutsche Bank is also experimenting with the concept. As one observer put it: 'no one thought they'd ever see Deutsche Bank next to the cheese counter!'

Major food retailers are at the forefront of these developments but over the years many different retail groups have shown the capacity to redevelop and renew themselves:

- Boots in the UK is a very different store in 1998 than it was in 1988. It had long been seen as an excellent but traditional pharmacy operation with an extended range in basic home and baby products. Since that time it has trialled and tested a number of different formats and has evolved into a total health and beauty care operation. It has a substantial amount of space dedicated to perfumery and hosts all the leading brands from Chanel through to Jean-Paul Gautier. It has muscled into Body Shop's territory with a very successful range of natural care products and in its key urban centre stores it often now dominates low-calorie sandwich sales for lunch times. With so much innovation it is not surprising to find that Boots has enjoyed record sales growth and persistently high margins over many years.

## How Can Retailers Respond?

- Taco Bell faced extensive competition in the late 1980s and embarked on a long search that would clearly distinguish its operation and ensure it remained attractive. It saw an opportunity to be both aggressive on price and on food quality and set about fundamentally restructuring its supply lines and restaurant operations in order to deliver those goals. It especially invested in order-processing systems so it could serve customers fast and provide a generally higher level of customer service. It too has seen consistent sales growth whilst rivals like McDonalds have stumbled.
- Both Comp USA and Home Depot looked to keep their retail concepts fresh and alive by building their staff's product knowledge, developing help teams and enquiry desks and making service the number one goal. They changed from outlets with racks of products to 'information stores' where consumers could go to educate themselves about the range and make the most informed purchase choice.

There are many other examples of successful retailer evolution responding to competition and changing demand patterns. Aside from the retailers mentioned, Sears in the USA, Next, Kohl the US clothing discounter, Radio Shack, Asda the UK supermarket chain and others have all made impressive recent improvements in their store operation and the proposition they make to their consumers.

*Continued pressure to innovate*

There is no doubt the pressure to continue to innovate will be maintained. Even without the advent of electronic commerce, Verdict – the UK consumer research house – was reporting in 1998 that consumers are generally disillusioned with the current retail scene. 'It's been somewhat samey for a number of years and consumers have become bored.' While some retailers are pushing out the boundaries many are risk averse:

> 'Retail brand identities need to be revamped . . . new ideas are needed to engage shoppers' attention . . . if retailers want success it's up to them to inject life back into their shops.'

But while retailers may have demonstrated capacity for renewal in the past they are being slow at pioneering change for the new electronic age. Particularly when it comes to facing up to electronic shopping, many retailers have remained ambivalent (see also Appendix 1). They are struggling to reconcile their physical estate with the demands and possible attractions of virtual selling and buying. As the following recent reports demonstrate, retailers at large have not yet grasped the electronic nettle:

- 'There's lots of trial and experimentation going on but scratch below the surface and we still find a mainly defensive strategy. In some cases retailers appear to be seriously under-estimating the inevitable growth of remote shopping and the key role which the Internet will play. A survey of 46 retailers across Europe shows that most still think it won't make a lot of difference to how they run their business.'

  (*Pira International*)

- 'Internet shopping is a small market at the moment and we are not aware of any retailers making money in it yet. Nevertheless there is clearly demand and that will grow spurred by a variety of forces ... It will establish a new distribution channel which will suck business – substantial business according to reputable forecasters – away from existing channels. That undermines the profitability and valuation of these channels. We believe the big weekly or monthly shop is at risk. Some commentators view the superstore as indestructible. But experts said that about the Titanic too! It's a logic that flies in the face of retail evolution. Every dominant format is ultimately replaced. We do not expect superstores to be any different.'

  (*Henderson Crosthwaite, retail analysts, 1997*)

*Changing skills and competences*

Of course it's not easy for physical retailers to embrace electronic distribution. They cannot simply walk away from their developed way of doing business. They have created in the past 50 years one of the most efficient business systems to be found in any industry sector. They have built outstanding core competencies especially in location selection and management, buying, logistics and IT systems. They have invested in adding more space and learnt how to squeeze up sales per sq. foot to very high levels. They have sunk billions into their real estate portfolios. They either hold the freehold or have long leases, often with up to 20–25 year commitments. They have no choice but to consider – at least as a first step – every possible angle to keep that real estate profitable, maintain at minimum current levels of store traffic and ensure they get their pay back on their capital outlay.

## How Can Retailers Respond?

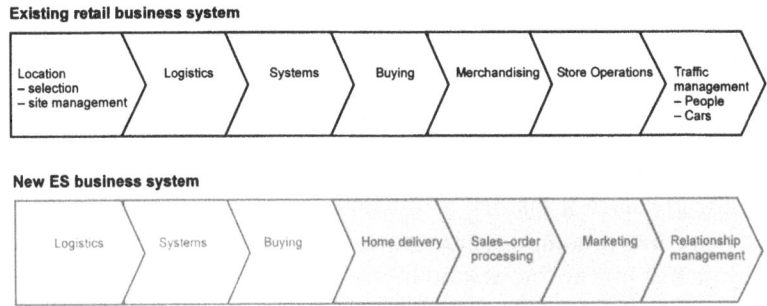

**Figure 6.1** Change in Business Systems

For established retail chains, electronic commerce has therefore initially been seen as more of a threat than an opportunity. It appears an alien world demanding new competencies and skills that they do not have, undermining their whole basic business system. When they do come to consider what will be required, it's clear that the new electronic business system for consumer selling is going to be significantly different from the current physical version (Figure 6.1). But only by confronting these changes and working out how to deal with them can they expect to turn these challenges into opportunities.

Some core competencies, such as location selection, will have limited application and relevance to a virtual world. Store operations and the management of that complexity may no longer be relevant if stores are unable to revitalise their proposition and consumers just stop visiting, ordering for delivery at home instead. New skills will be demanded – for example, in timely and efficient home delivery, handling electronic buying orders many of which may come through at 11 o'clock at night for next-day delivery in the morning. Marketing will take on a whole new significance as would-be electronic retailers learn how to reach and engage their customers and make their brand names relevant and fashionable for the new age. Relationship-management skills will be at a premium trying to forge new links with more remote consumers now with more choice, shopping flexibility and ability to compare prices.

*Making the transition*

Can retailers make this transition? Can they successfully migrate from a long-established way of doing business to something quite different? To what extent does their history of innovation and successfully responding to and shaping consumer shopping habits give them the courage and confidence to face up to these new changes? How long have they got to get their minds around the various options? Is there enough time left to start the planning and development so they are prepared for the critical mass point of 2005?

Certainly retailers don't have to make this transition on their own. They already have a history of outsourcing key activities. In the early 1980s there was a boom for transportation and warehousing contractors as they picked up enormous amounts of business from the major retailers either taking over existing retailer activities or building new sites and operations. Since that time outsourcing has become more sophisticated and many retailers have established joint ventures or partnerships especially in 'back-office' areas. IT has often been handed over completely to third parties, others have established large-scale call centres dealing with burgeoning phone/fax order operations, credit queries, invoice processing and in banking for example actually pioneering customer orders and confirming transactions.

There is no shortage of companies willing to offer themselves as partners to retail operations. Companies in the UK like Sitel and BT Connections offer a range of call-centre activities. Global organisations like EDS, CSC and Andersen are all moving to offer a range of outsourcing and support functions. On the home delivery side, companies like Northern Foods in the UK are considering how to exploit their long-established milk-delivery operation to individual homes to take on a range of broader products and services (the milk round becomes the daily or weekly food delivery service?) Sears in the USA has pioneered a whole series of home help/home service functions and has such a solid reputation with consumers that it is in a strong position to move into the general home delivery business for others. To the extent that consumers are reluctant to allow companies to bring goods into their home wouldn't tried and trusted Sears uniformed personnel have a much warmer reception?

Organisations like UPS and Fedex are themselves extremely well placed to take advantage of this situation with sophisticated parcel pick up, tracking and residential as well as office delivery systems. Other third-party providers who might get involved in this arena include the big mail-order houses. For example, GUS and Littlewoods in the UK have highly advanced telephone-ordering, credit-checking, order-processing and home delivery infrastructures. They could partner with retailers looking to embrace and come to terms with the new electronic business systems

## How Can Retailers Respond?

**Figure 6.2** Value Networks

and the skills that would be required. Alternatively they could leverage their *own* home delivery competencies and like Peapod, Shoplink, Flanagans, Food Ferry and others establish their own food and other products' home delivery operation.

Retailers need to examine this new challenge and perhaps adopt a new kind of operating model that has been pioneered by CSC, Professor Michael Porter and others – the value network (Figure 6.2). This would put the retailer at the hub of a total solutions package. It would enable a company to provide a major response to the electronic challenges but do so in a highly leveraged partnership vehicle, bringing in a range of partners and outsourcers to provide the best expertise.

Operationally there is no doubt that retailers could achieve great success as pioneers and major distributors in an electronic environment. They have years of retail experience. They are well placed in the present value chain. And they can enhance their market position and capabilities in the new era by 'value networking'. They can bring a select range of partners and experts to help them realise their goals.

Their biggest obstacle is in their own heads. They are struggling to deal with the apparent channel conflicts. Is there an effective marriage out there that enables them to retain the viability and profitability of their current real estate while at the same time embracing the new electronic world? If there isn't then how can they migrate from physical to virtual selling without losing their customers and in the doom scenario see profits collapse before the electronic operation is up and running?

There is detailed discussion about managing channel conflict in Chapter 12 but it is worth emphasising right now that the transition *can* be accomplished. There are a number of 'success models' out there that point the way and demonstrate the various options. The key lies again in segmentation. There is no substitute for really understanding who are the different customer types, how many want to shop electronically and how many do not, how many are undecided and what will drive their decision-making, whether apparent interest in the electronic world is in fact relatively superficial and whether they can be kept visiting the stores if the stores can respond enough to them. The ES Test and the analyses that go with it will help determine the most appropriate response to each segment group.

Retailers have such an outstanding platform on which to build. They have great brand names, tremendous reputations with their customers, strong and often loyal followings – they are part of the fabric of society. That established position will help them withstand the market place changes we have described, but it provides no guarantee into perpetuity that business will keep coming through their door. They can change and are well placed to change, but the planning needs to start soon.

**Future face of shopping**

As retailers face up to shopping in the twenty-first century what type of environment will they be dealing with? Will there still be the current mix of giant out-of-town shopping centres and malls, large inner-city centres, small town high streets, neighbourhood stores and 7–11-type convenience outlets? Will all parts of this retail scene still attract significant customer flow and continue to be attractive and profitable? Will some shopping areas be more impacted by electronic commerce than others? Which will decline and which will gain?

Examining the trends, it seems clear the future face of shopping will experience the different forces and dynamics of the electronic age and these are bound to have a significant effect. When people do decide to go out shopping, as they surely will, many will go out with different expectations and needs:

- basic groceries and other regularly purchased items will often already have been delivered to their homes
- other products or services which in themselves have a high primary appeal to 'virtual' senses of sight, sound and intellect, will have been prescreened and ordered up electronically from the home PC or TV:
    – products such as books, music, video, computer software and hardware
    – services such as banking, travel and holidays, real estate searches

## How Can Retailers Respond?

Consumers' shopping needs will therefore take on a different emphasis. The focus will be on experiencing, seeing and trying *new* products and services rather than filling the shopping cart with the weekly shop. Some of these trends and changes are not new. Even before the advent of electronic commerce, retailers have been trying to make the whole shopping experience more fun and exciting. They have been developing new formats and searching for ways to reinvigorate their physical space:

- the large supermarkets are adding a range of services like banking, pharmacies, coffee shops to provide a stronger 'one-stop shop' environment
- big new centres are being created typically edge of or out of town which have integrated social and leisure experiences with movie centres, choice of restaurants, fitness and sport facilities – all designed to create an extended stay and family experience.

As these new, more exciting larger centres have opened so small strips and high streets or small groups of neighbourhood stores – the mom's and pop's – have been facing increasing pressure. Consumers are more naturally gravitating to the new shopping experiences (Figure 6.3). They are leaving behind the centres that are old and tired or just too small to provide for what they need. As a result many small-scale shopping areas are under increasing threat. Family-run independent outlets or small chains are suffering as the large centre category killers like Toys R Us or

**Figure 6.3** Share of Retail Sales, by Location (UK): Out-of-Town/Regional Malls vs High Street

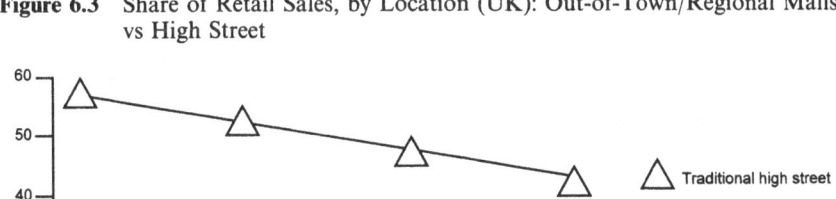

*Source*: Corporate Intelligence on Retailing.

Home Depot have opened and immediately penetrated through a wide catchment area with their greater range and choice. Smaller bank branches are also closing, taking away what is often seen as a commercial cornerstone of any flourishing shopping environment.

Combine those trends now with the impact of electronic commerce and there's bound to be accelerated change in the physical retail scene. Small and mid-sized retailers in small and mid-sized centres are likely to see intensified sales and profit pressures. As we move into the twenty-first century, there will be a growing trend toward shopping as more of a day-out activity. There will be even greater interest in the giant centres where the family can spend an extended period looking at new products, trying things on, buying special presents for gifts and birthdays and where the kids also can enjoy themselves.

There will, however, be some exceptions to this. Individual outlets or shops, despite being small in scale, will continue to flourish where they have developed a special appeal. This could include tourist centres e.g. the Cotswold village shopping street or Fisherman's Wharf in San Francisco. It could also include centres that feature destination shops such as specialist antique shops or factory shops where visitors can walk around a factory seeing glass being blown or chocolate made and then buy samples or gifts afterwards. Also likely to survive will be the convenience 7–11s. People will always run out of milk, bread and beer or forget certain items or need something at the last minute.

What will emerge will be a polarisation of the retail shopping scene (Figure 6.4). The larger centres will grow in attraction and the small shops and small centres will begin to die out. There will be a snowball effect, too. The less successful the small centres become, the more outlets will close, the less attractive the surviving businesses will be. And these changes are not that far away. The research is clearly pointing to around the middle of the decade as the time when electronic technology, communication and infrastructure will mature sufficiently to satisfactorily meet and encourage this polarising consumer demand.

Electronic shock! A revolution in the shopping environment is about to take place. It won't immediately affect all consumers. It won't immediately impact all retailers. The USA is ahead of western Europe, other nations are still in the process of waking up to the challenge. Not every mid-sized high street or mall will collapse in the next decade. Some will prove more resilient. Some will have a catchment area of more traditional social shoppers. Others will not have a major mall or centre that is accessible enough to seriously undermine their business. But the trends are firmly in place, the forces for change are gathering momentum

**Figure 6.4** The Shopping Environment 2005 and Beyond

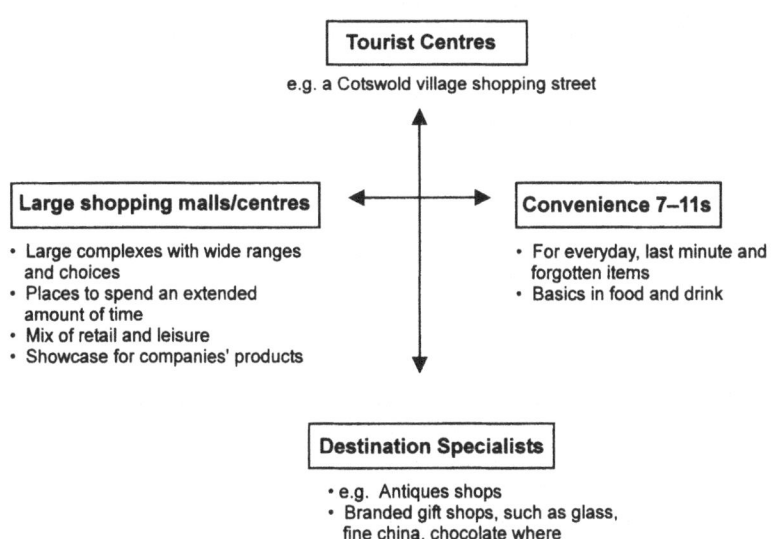

and there is a certain inevitability about the transformation that will occur.

Doing nothing right now may still seem a viable option because there are still many sceptics and the market change is still relatively small. But it is growing at an exponential rate. What must be concerning for all retail groups is the head of steam that is building up among the computer, telecommunication and content suppliers especially who are determined to turn business and commerce electronic as quickly as possible. Resisting that and ignoring underlying consumer demand does not appear a viable option. A response is required and the next chapter lays out the ten alternatives that could be pursued.

# 7 Ten Strategic Options for Retailers

If the ES Test shows that a retailer's products or services have some potential to be purchased electronically then how can that retailer respond, what options are there?

In working with companies in this area research and investigation has identified ten alternative strategies or response options. They cover a range of initiatives from taking just a few small steps to acknowledge the ES interest, through to the most radical involving a full switch of the business away from real estate to virtual shopping.

The ten strategic options are:

1. Information only
2. Export
3. Subsume into existing business
4. Treat as another channel
5. Set up as separate business
6. Pursue on all fronts
7. Mixed system
8. Switch fully
9. 'Best of both'
10. Revitalise and buck the trend.

Before we get to the definition and illustration of each of these strategic options, it needs to be emphasised that they don't all require the abandonment of the current retail estate. Even once the consumer interest and need is sufficiently understood and defined there is still a choice. There is still the option, for example, of 'bucking the trend'. Retailers can choose to review whether the real estate can somehow be so revitalised or developed that customers will still want to visit and shop no matter what the alternative ES temptations. Even moving down the ES path there are many stages of evolution and a number of the options do assume that the physical space and the stores can still play a viable and vital role.

**Figure 7.1** What Path to Take?

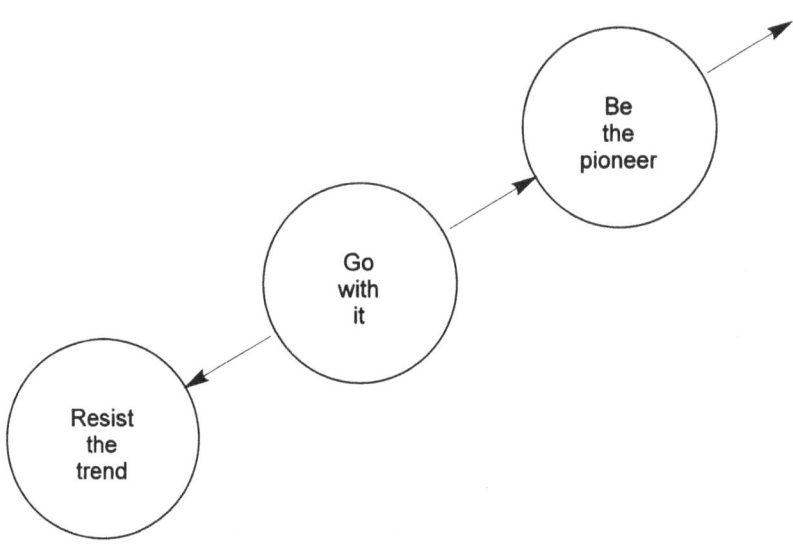

As with any strategic decision-making an organisation of course has 'free will'! It must understand its markets but need not be a slave to them. It has the opportunity to influence and shape future market conditions as much as be influenced and shaped by them. Large established retail operations like Marks & Spencer, Tesco and Sainsbury in UK and Wal-Mart, Home Depot and Brooks Brothers in the USA have been shaping the retail scene about them for many years and consumer shopping patterns are partly driven by the initiatives and changes these companies have made. So existing retail groups must be prepared to think through all the options and determine whether they want to try to resist the trend, go with it or be the pioneer (Figure 7.1).

Taking the ES Test is an important first step in deciding what path to take. But as it's still early days the test may show consumer interest still too indeterminate and uninformed. In such circumstances the middle route of 'going with the flow' is a viable approach – hedging bets till more certainty can be identified. But even that needs to be a proactive deliberate decision not passive inaction. It needs to include a rigorous and continuous monitoring of the market place to check when, if at all, a more forceful strategy is required.

Whichever route is chosen, as we shall examine in Chapters 14 and 15, excellence in execution will be critical. This is especially true if there is a determination to resist the trend and keep consumers coming to the store.

This cannot be emphasised too highly. As with any major market evolution the winners that we all come to talk about are those who make an exceptional commitment to ensure their chosen path works. This requires clear strategic thinking, stark milestones and targets, every $ of investment and resource made available to fully fund and staff up the initiatives and a communication and involvement of the work force that delivers an excitement and identification with the strategy and gets employees to go the extra mile in what they do and how they meet and serve their customers. Now more than ever, achieving total customer satisfaction with the shopping experience – however provided – will be a litmus test determining success.

By way of definition, options 1–8 (Table 7.1) assume increasing levels of consumer interest in ES and, with that, greater movement away from existing physical real estate toward an electronic environment. Options 9 and 10 are somewhat different. They both 'buck the trend', seeking ways to defend but also exploit the existing real estate and retail core competencies.

We can illustrate each of these options:

1. **'Information only'** – This acknowledges there will be some ES interest and some exploration by target consumers of what's available electronically in most every business sector. For this reason some presence at minimum is required, as much to respond to this consumer exploration as to protect turf against competitors taking a more aggressive approach to electronic shopping.

Table 7.1 Ten Strategic Options

| Strategic options | ES responsiveness | |
|---|---|---|
| 1. Information only | ↓ | ↓ |
| 2. Export | | |
| 3. Subsume into existing business | Stronger the consumer interest in ES | Evolving away from physical retail base |
| 4. Treat as another channel | | |
| 5. Set up as a separate business | | |
| 6. Pursue on all fronts | | |
| 7. Mixed system | | |
| 8. Switch fully | ↓ | ↓ |
| 9. 'Best of both' | ↑ | ↑ |
| 10. Revitalise | | |

This is the state which most retailers and indeed most organisations of any substance have reached at the current time. Web sites largely are information-giving, indicating what's available but often referring the would-be customer to a list of existing retail outlets or other established means of purchasing.

For example, UK retailer B&Q – a leading DIY retailer – has a web site which at this stage deliberately does not sell any product. Instead it focuses on promoting what B&Q offers in terms of products and services and then tells the customer *where* they can be obtained. B&Q has tried to protect this by getting the right to use the generic web site address: www.diy.co.uk so any customer exploring the sector will first reach B&Q and then be told where and how to purchase. Such 'control' of web site exploration could potentially cut out competitors who may have a much greater transaction-driven web site but may never in practice be reached.

This B&Q example provides a good illustration of how a retailer can manage the ES potential around its products and services without potentially needing to change its retail operation. Though of course there's a point to watch and monitor for where ES interest becomes so strong that the absence of transaction facilities could actually turn the consumer away determined to find the site or electronic means that they can use to shop.

2. **'Export'** – This approach looks to ring-fence existing domestic retail establishments but use ES interest to gain access to new markets and new customers.

For example, Blackwells is a leading UK bookseller with 82 outlets in university towns and campuses across the country focusing on academic books. For the past 100 years it has had a mail-order service offering its books to wholesalers who will in turn sell them on to individual bookshops in overseas locations or to colleges and universities.

Now with the new web technology Blackwells is in the process of reinvigorating its entire export operation offering books not just to wholesalers but also direct to consumers. Borrowing from some of the experiences and learning at Amazon.com, Blackwells has given its site and its transaction-processing capabilities a multi-million-dollar revamp. There is a full listing of over 1 mn specialist and academic titles, reviews and recommendations, delivery anywhere in the world, on-line order tracking and account management.

To book buyers outside the UK Blackwells is to all intents and purposes a dedicated on-line bookseller. To UK book buyers, while the on-line facility is still available, Blackwell's current strengths are its convenient retail locations close to places of learning offering a specialist destination shop.

3. **'Subsume into existing business'** – Like the previous two options this does not seek to change the existing retail operations. However it acknowledges that there will be some ES interest but that interest will, so far as it can, be subsumed or integrated within the existing retail establishment.

Safeway and Sainsbury are two leading UK grocery food retailers who have pioneered this approach – unlike one of their major rivals Tesco whom we shall investigate later. Each has established 'Order & Collect' or 'Collect & Go' operations:

- 'Since end of the year, we have announced an initiative in home shopping. As many customers find shopping for bulk commodity items both tedious and time-wasting, Collect & Go enables ABC customers to order these products by phone or fax from a personalised shopping list based on their previous purchases.'

*(Safeway 1997 Annual Report)*

First, note Safeway's recognition that 'customers find shopping [for certain items] tedious and time-wasting'. But Safeway still want their customers to visit the store. Rather than 'enabling' them to stay away with a delivery to the home service, the current strategy requires customers to come into the store even if it's just to visit the collection pick-up point. The aim is to get 'electronic customers' to still shop and walk the aisles for fresh foods and meats and other special items even if they are getting the basic commodities picked for them

The Safeway approach is built around an initial 'personalised' shopping trip where the customer together with a Safeway service representative walk the aisles establishing the 'personalised shopping list'. Once in the Safeway computer a phone call and a reference number from the customer will trigger a Safeway shop assistant into picking the order from the aisles on the customer's behalf so it will be waiting for collection at a set time.

There are executional issues still being sorted out such as managing the personalised picking and collection times so they don't bunch up at the same time of day and dealing with the common issue of substitution – what if the preferred brand or size or variant is not in stock that day? But in this period of trialling and experimentation Safeway is not afraid of making the occasional mistake if it will help to learn how to make the operation a success.

4. **'Treat as another channel'** – This is currently a common approach and many retailers are seeing ES as simply an alternative means to reach a target group of customers.

UK retailers like Tesco, Victoria Wine (part of Allied Domecq), and Dixons are looking to exploit the new technologies in this way. Their aim is either to offer greater convenience to existing customers or to attract new customers who can't reach their stores or typically shop elsewhere. For these companies electronic selling has not yet achieved the status of being a totally separate funded and resourced business. It still shares overhead, systems, back-office functions and personnel with all the other main business activities. It's treated in the same way as other channel experiments in the past such as with mail-order or franchising. The company hasn't yet got serious enough about it to set it up in its own right.

Part of this 'hedge our bets' experimental approach has also seen some retailers participate in others' web sites rather than leading with their own. So some can more easily be located by selecting the commercial service provider's mall. AOL and CompuServe each offer this service for an annual fee plus a percentage of the sales. Here it is the provider not the retailer who typically takes the lead in design and promoting the product offering and directing consumer traffic to the site and the virtual shops on it. CompuServe's mall included for example major retailers such as JC Penney, Brooks Brothers and Virgin music but also smaller groups such as Health & Vitamin Express, the Futon Shop and Access Cameras who can reach wider audiences that way. AOL has FAO Schwartz, Eddie Bauer and Godiva chocolates. A further alternative is the host-mall arrangement which is to participate in an open Internet web mall site such as NetMarket, Shopping.com or the Internet Mall. Each of these 'hosts' can again take the lead in establishing the site.

Some retailers will be experimenting with both their own web site as well as participating in on-line shopping malls. So Sears Roebuck, for example, has a presence in the Prodigy mall, in an open web mall site and has its own home pages. Equally UK's Barclay Square hosts sites for a number of retail groups who also operate on their own, such as Argos, Thomas Cook travel and Toys R Us.

5. **'Set up as a separate business'** – This is an important evolution from the previous approach. It's a sign that the opportunity is seen as

stand alone and significant in its own right. It also recognises that it is a 'different business' with different skills and competencies required to make it work.

An analogy can be drawn here with Royal Bank of Scotland's (RBS) involvement with Direct Line insurance. Unlike other financial institutions, RBS set Direct Line up with its own management team, funding and operation and gave it its head. It did not seek to limit or control what Direct Line did and was prepared to live with any cannibalisation that arose for its existing businesses.

Given the general immaturity in electronic shopping at this time, there are few examples yet of retailers moving down this path. However Tesco's own direct business has a significant infrastructure of its own; Next retail has made a tremendous success of its separate mail-order business Next Directory, which works well in reinforcing the retail brand. Marks & Spencer too expects the growth in its own mail-order catalog business to be accompanied by significant improvements and investments in infrastructure and mail-order competencies.

No doubt as we move through this current era of experimentation and trial more retailers will bite the bullet in this direction, though the cannibalisation issue is likely to be closely measured and monitored.

6. **'Pursue on all fronts'** – Some retailers are determined to pursue every channel that's open to them. They want to reach all their possible target consumers, be as competitive as possible and not allow others to steal a march in a market channel area.

Eddie Bauer, for example, sells its casual men's and women's clothing through some 500 stores in the USA, Germany, the UK and Japan. It also has a developed mail-order catalog operation (to be expected given its parent Otto Versand's skill base) and is currently exploring on-line sales on the Internet via its own web site. The company has no particular prioritisation between these three channels but simply sees all three as available and necessary routes to market. If there is any cannibalisation, it's as yet immaterial as the company seeks to expand globally racing to reach as many new customers as funds and resources allow.

Great Universal Stores (GUS) is the UK's largest mail-order company and has a highly developed order-processing call centre and home delivery service. GUS also houses well known retail brand names like Scotch House and Burberrys which can be purchased mail-order, retail and now on the Internet through the Shoppers Universe retail site.

Levi's represents a slightly different illustration of 'pursuit on all fronts'. While still primarily a manufacturer and distributor via wholesalers, Levi's has now developed a large number of its own retail sites with some 32 in the UK alone. While living in constant conflict with other retailers, Levi's has remained undeterred and determined to explore all channels to market. It too now has its own web site which strongly reflects the life-style image and presentation seen in its store outlets, with games, questionnaires and related features and information. While at the time of writing this web site was only for information and entertainment purposes, the understanding is that plans are being developed to enable customers confident in their size and style requirements to order electronically.

From the financial services sector, National Westminster Bank offers a good illustration of a 'pursue on all fronts' multichannel player. Figure 7.2 sets out how they see their business.

**Figure 7.2** National Westminster Bank: Multichannel Operator

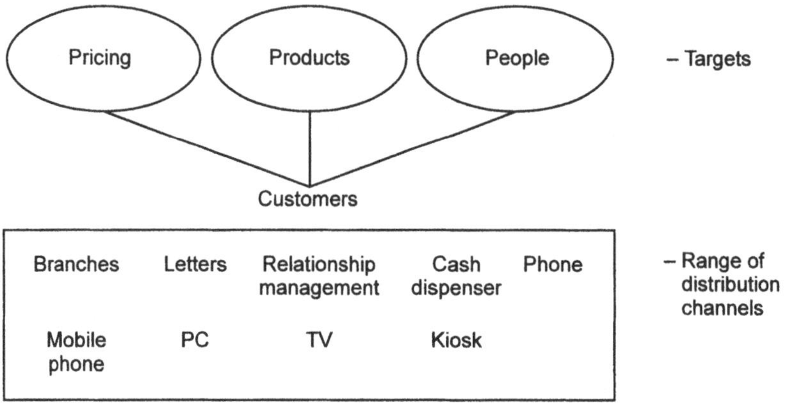

```
   ( Pricing )   ( Products )   ( People )       – Targets

                  Customers

   Branches   Letters   Relationship    Cash        Phone      – Range of
                        management      dispenser                distribution
                                                                 channels
   Mobile     PC        TV              Kiosk
   phone
```

7. **'Mixed system'** – This strategy goes a long way in addressing the importance of ES but also recognises that physical retail shopping will not disappear, that there will be key malls and shopping centres where a continued presence is vital. This approach could also be called 'Flagship stores + On-line delivery'.

As described at the end of Chapter 6, the future shopping scene does not mean the elimination of physical retail shopping. People will still go out to shop but it will be more of a day-out leisure activity as opposed to the chore job of getting in the basics. In this scenario the big malls and shopping centres will stay strong and become shopping leisure entertainment meccas which will reach out into wider catchment areas. In these centres, consumers will expect all the main retailers to be still represented.

Retailers will want to be there too because the stores in those centres are likely to both attract a lot of traffic flow and be profitable. But also these stores will act as flagships and as major promoters of the retail brand name. It will be this brand reinforcement that will help build and sustain the complementary on-line direct electronic selling operation (Figure 7.3). This approach becomes a twenty-first-century version of hub and spoke with the flagships operating as key hubs supported by a few central warehouses supplying the stores and reaching out like spokes to support an 'on-line + home delivery' network across the country.

The challenge will be to make the capital flagships a total experience in the product and its life-style. Levi's, Doc Martens, DKNY, Disney, Warner have already established these major flagship showrooms. Tower Records currently has 7 major 'flagship' stores across the UK, its largest in Piccadilly Circus is 37,000 sq. feet. Tower is currently building its on-line operation to reach the many consumers who cannot get to one of its main stores. It's taking maximum advantage of the multimedia to also offer its vast catalogs electronically and has formed alliances with other commercial on-line networks so it has a highly visible, well advertised web presence.

**Figure 7.3** Flagship and On-Line 'Mixed System' – US Example

Tower does not intend to build physical stores across the country. It sees its future around flagships to promote the brand and capture a share of retail sales *plus* electronic selling and home delivery to reach out to the whole population.

Virgin Megastore is another example in the same product category. Indeed Virgin's heritage was originally in mail-order, then branched out to its own retail outlets. Now it is building major flagships in capital cities and other large conurbations in the key shopping centres and is looking to complement that by rejuvenating its old mail-order business and turning it into a more aggressive on-line selling operation.

This mixed system approach recognises that strong brands will be key in a virtual shopping environment. It works on the premise that electronic brand strength needs to be showcased and reinforced through a select store network. Established retail operations already have a major advantage selling in cyberspace as their brand name and reputation is widespread. But as ES shopping picks up and there are fewer visits to the store, organisations are going to have to find smarter ways to keep the brand name and awareness high. Part of this effort will go into various forms of advertising and prompting on others' web sites and we shall explore this later. Part of the effort too can go into the retail operations themselves so that the visit to the shop becomes a true 'flagship' experience, more than just walking the aisles or looking through shelves or racks of products.

8. **'Switch fully'** – Will any retail operation be bold enough to switch fully, to take the path of gradually shifting away from its physical sites and becoming a dedicated electronic trader? Certainly there are only a few examples of this happening so far. For example British Airways has been steadily cutting back on its physical travel shops as its electronic operation grows and as Microsoft's Expedia continues to demonstrate the ease of on-line ticket buying. More dramatically, UK travel group, the AA, announced the closure of all its 142 high street shops saying: 'people now prefer to do business over the telephone or web site'. As the virtual market develops might we see other retail operations or those whose products/services most lend themselves to ES gradually making the transition? Are not retail banks prime candidates (see also Appendix 2)? The branches are becoming increasingly redundant as ATMs, debit and credit cards take over the basic cash and cheque operation, telephone and PC banking continue to develop and consumers are getting increasingly used to maintaining their bank account without ever visiting a

branch. At some point as electronic banking grows the economic viability of each and every physical branch will come under greater scrutiny. How many, if any, will survive? Some banks and building societies like Nationwide in the UK do seem to have already embarked on a deliberate strategy to get customers to gradually deal with them electronically. As the switch occurs they are closing branches accordingly.

In the same context, retail chains like Blockbuster Video will increasingly see their product made available through satellite, digital TV, through cable or over the Net. How long again before the number of visits to the physical store so reduce that it simply becomes uneconomic to keep it open? In such circumstances won't these retailers be forced to 'switch fully' or simply shut down?

Argos, the UK retail and catalogue mail-order business (now being merged with GUS), is one operation that is aggressively challenging its own store operations. Its home delivery operation Argos Direct saw sales up 42% in 1996 to £100 million. It also has a growing Call and Collect operation with ten stores open taking orders by phone and fax or catalogue with collection next day. These stores are small compact units of some 1500 sq. feet and if they are successful in their own right Argos will surely be examining how many flagship superstores it needs showcasing the entire product range. Certainly it is unlikely to need the 300 or so stores for every town in the UK that it currently has today.

For greenfield new entrants, of course, the choice is easier. The lower investment and set-up costs, the immediate global reach, the recognition of which products are most ES-sensitive means that the past two–three years has seen a number of new companies opening straight onto the Net with an exclusive on-line offering. Companies like Amazon.com, Carshop, CD Now, The Fresh Food Company, Peapod, Internet Bookshop, Interactive Music and Video Shop, Orchard Direct, 1-800-FLOWERS have all set up dedicated electronic sales and home delivery operations which are slowly establishing themselves. Unencumbered by existing physical sites and with a lower cost base these organisations are learning how to take advantage of their position in the market. In certain sectors such as music, books, computers and financial services these electronic operations are already seriously challenging their established physical retail competitors.

9. **'Best of both'** – Is there an approach which can meet the electronic shoppers' needs but still retain the integrity of the existing retail

operation? Is there a response which allows retailers to sustain their physical retail operations yet still takes them forward into the twenty-first century – a development path that can achieve excellence on both the physical *and* virtual dimensions?

The 'mixed system' – option 7 – goes some way in this direction but it assumes reducing to a few flagship stores supporting and stimulating a far-reaching electronic sales network. 'Best of both', in contrast, assumes that most if not all existing stores *can* be retained. It also assumes this is done *alongside* a developed electronic operation.

The essence of this approach is the physical shell of the retail outlet. Let us assume we have this sunk investment and our challenge is to move it on so it meets the new shopping environment in 2005 and beyond. What changes would we make, how would we lay out the store? Would we continue to have rows of aisles and racks and shelves or should we assume a number of our customers will already have got the 'chore buying' – the basics or the regular purchases – out of the way before they visit. Let's imagine we're a food grocery store like a Tesco, a Sainsbury, a Giant or a Kroger's Supermarket. We have all this space, now what do we do with it?

We want to achieve a store for the future, a place people will want to visit, a destination shop for the twenty-first century. We want to create an environment which keeps attracting customers and if possible gets them to stay longer, spend more money in total and enjoy the total store experience. The basic and regular purchase items must be dealt with as efficiently and quickly as possible – get them out of the way – bring out the fun side of shopping. There will always need to be a certain functionality, but can shopping now also be made pleasant and relaxing? For example, the store might feature an extended coffee shop and lounge. That could be a place where people can meet, socialise, relax *but also* place their shopping orders. They could shop conveniently and electronically via touch screens on the side of their table. Payment made there and then, so no check-out counters or long waits standing in line, no wheeling out unwieldy shopping carts to the parking lot. Simply drive out via the Collect & Go station where an assistant puts the bags in the back.

In the next chapter, a possible store of the future is described and illustrated in more detail. In planning its development retailers will need to review the combination of products, services, service and environment to find the best combination for their target consumers. They will need to test and trial what works and what doesn't and how

## Ten Strategic Options for Retailers

**Figure 7.4** The 'Best of Both' Shopping Experience

each can best interact to serve up a totally satisfying 'best of both' shopping experience (Figure 7.4).

Ambitious retailers will want to investigate opportunities along each of these four dimensions in defining their store of the future. What level of individual customer service and services will be appropriate, what new initiatives in relationship management, product knowledge, child care, and order-processing systems will be needed to ensure that the customer experience is fulfilled and not frustrated? What environment will need to be created? Does it still look like a Tesco or a Wal-Mart late 1990s vintage? What's the buzz that lifts it into a desirable shopping experience? Is there sufficient excitement simply in the new features to the store and its new layout? What's the opportunity, for example, in genuinely creating a community centre where people meet and pick up their shopping later? Can that lead to the development of a strong catchment area identity and 'community spirit'? How might that be captured in-store? Does the store become like the church hall with local news, local events and meetings? Will that translate into increased consumer traffic? And where is the line drawn on the product offerings? What is showcased and what is relegated to the back of store-picking warehouse? Does it become a negotiating issue where manufacturer/suppliers can only get products into the store front on special payments and promotion terms?

Some of the challenges for building this 'store of the future' include investigating whether there is enough space – is the existing shell big enough? Which mix of services is most appropriate and if there's not room for them all then which ones have most appeal? Also for consideration will be the pace of the changes. Do stores transition in short order or do they evolve to move gradually to this new environment? Electronic system costs may be substantial, store labour may be more intensive and the economics of the store will require careful consideration to retain satisfactory levels of profitability.

As retailers consider the options there are clearly many questions to be addressed in understanding what can be achieved, what's profitable and what isn't and what will best meet consumer demand. Certainly the big food grocery chains all have large enough stores and a sufficiently aggressive new store-opening plan that in the next few years 'stores of the future' can be developed and trialled to find out what works and what doesn't. The trick will be to set out this new store at a point in time when consumer interest in ES is clearly on the increase, when technology is more readily available to make the electronic purchase experience fun and easy, and to choose a geographical area where consumers are more likely to be responsive to these sort of innovations.

Of course it requires a certain boldness to carry out this sort of store trial and development. Some of the bigger groups are in fact not that far away today from this type of set-up, with experimental 'Collect & Go' pick-up points and additional services being brought into the store. But it is the ones who continue to experiment boldly who will have the best chance of shaping and influencing how the market place develops and stealing a march on their competition.

10. **'Revitalise and buck the trend'** – This option makes no concession to the forces of electronic retailing. It takes a determinedly defensive stance toward the existing physical retail operation but looks to revitalise it so consumers want to come and continue to shop just as they did before. It assumes changes in the retail format, but unlike the previous strategy these are purposely not aimed at meeting the electronic shopper.

A likely path for retailers choosing this option is to turn their retail operation into a Retail + Leisure environment where visiting the store becomes a more rounded fuller experience. 'One view is that many retailers will also have to become entertainers,' observed the *Financial Times* in a recent survey, 'as internet shopping grows consumers may only be tempted into the shops if they can expect an environment which majors on fun.' And there are two ways to achieve this. Either each store works out its own individual solution, or groups of stores in a mall, a shopping centre or even in a high street combine and develop a total shopping and leisure solution which all can benefit from but is bigger and better for having the funding and support of a large number of different businesses.

In response there is a wealth of ideas being trialled around the world, ranging from movieplexes to a full beach and kayak racing inside West Edmonton (Alberta Canada) mall. There are also dozens of shopping centres now boasting fairground entertainment including one in Durban, South Africa which has a ferris wheel protruding through the roof of the centre. Restaurant and food courts abound and developers are now pushing out the boundaries to explore what leisure facilities will give their new shopping centre that buzz and excitement that will generate the desired traffic flow.

As electronic shopping takes off, those determined to buck the trends will face growing pressure to find the right blend of leisure and retail. Once established it will then require continued monitoring to ensure the leisure retail concept remains up to date and endures and there is the right mix of retail tenants to complement the leisure theme. Might we begin to see retail/leisure parks targeted at different customer groups, with some aiming at the younger shopper while others target the older more sophisticated shopper whose concept of leisure is likely to be less noisy and helter skelter and geared to more sedate and subtle interests?

Achieving the appropriate retail and leisure state will require significant investment and proactivity on the part of a retail chain. It's likely to be something retailers will want to seek experienced leisure operators as partners to both share the risk and capital but also to access the core skills and competencies required to make the leisure initiative truly work and pay back.

\* \* \*

**Summary**

To help evaluate and compare these ten strategic responses let us draw out a couple of summary frameworks. First we can contrast the risks and potential of each (Figure 7.5).

Each option has its own risk:reward profile and will have a different appeal to each retail group depending on that group's own culture, its ambitions and its desire to resist the trend or be a pioneer.

118

**Figure 7.5** Comparing the Options

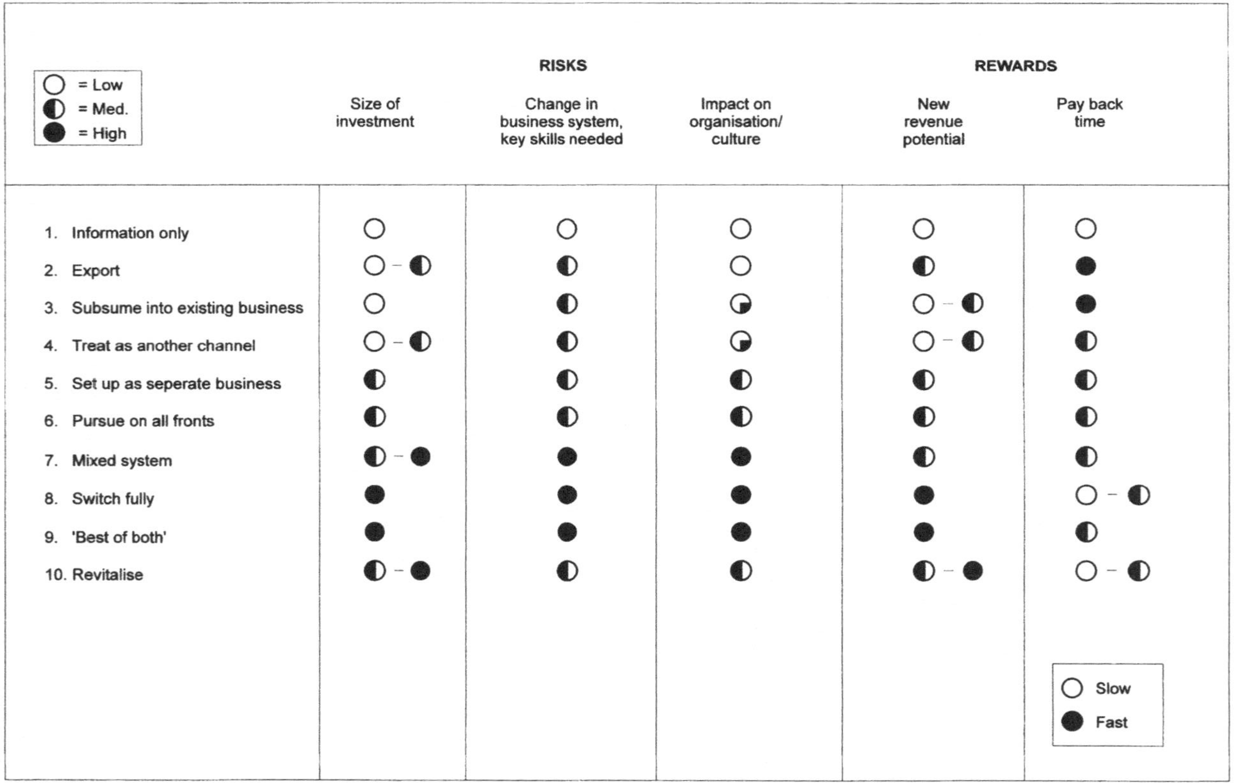

## Ten Strategic Options for Retailers 119

**Figure 7.6** Choosing from the Ten Strategic Options

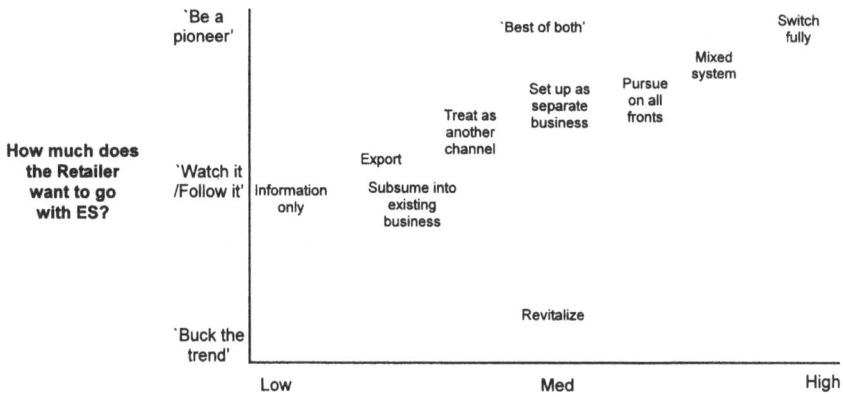

We can also turn back to the ES Test for help in choosing which of these options is the most relevant to the specific product/consumer mix that an individual retailer is dealing with (Figure 7.6).

Now is the perfect time to be considering these ES challenges and determining the response that will prevail in the new market environment. Critical mass in electronic shopping is still some years away but the opportunity has arisen to evaluate and plan today, taking into account how the market will likely develop, using the ES Test to get a handle on specific target customer needs and determining what risk and reward ambitions are most attractive.

But this decision-making cannot be delayed much longer. 2005 is racing ever nearer in planning and investment terms. In the meantime new store openings are already being scheduled, land is being purchased and designs and formats defined. The new order in the next decade will not allow 1980s formats to continue to survive and succeed. Changes and redesigns can be thought through now enabling the next generation of stores to be built – or the new world of electronic shopping to be created.

# At Sainsbury's Self-service shopping is EASY and QUICK

1—As you go in you are given a special wire basket for your purchases.

2—The prices and weight of all goods are clearly marked. You just take what you want.

3—Are you a fast shopper or a slow? You can be either when you shop at Sainsbury's!

4—Dairy produce, cooked meats, pies, sausages, bacon, poultry, rabbits and cheese—all hygienically packed.

5—Meat is served from Sainsbury's special refrigerated counters. Or you can serve yourself from the cabinets.

6—Pay as you go out. The assistant puts what you have bought into your own basket and gives you a receipt.

Sainsbury's Self-service
© J. Sainsbury plc

## Shopping in the electronic age

1. Shopping from the comfort of home

2. While the store staff pick your order . . . .

3. . . . . you can be out enjoying yourself

4. Your shopping's delivered when you want it

Iceland's home delivery service

Ordering your Tesco shopping on the Internet

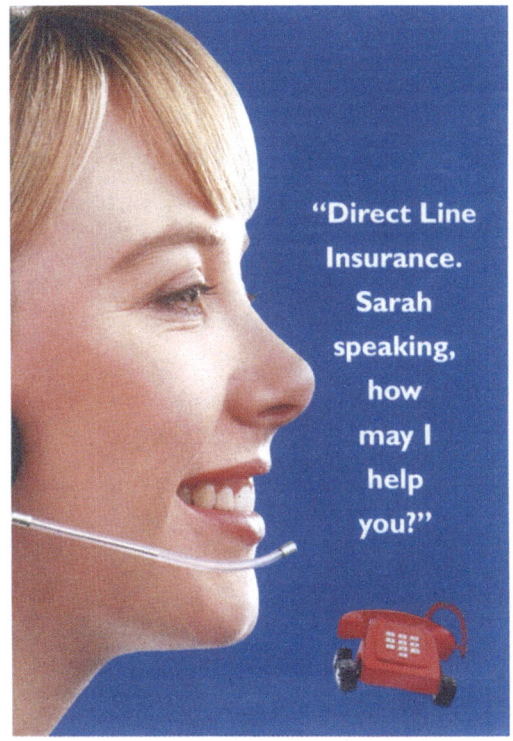
Direct Line – insurance over the phone

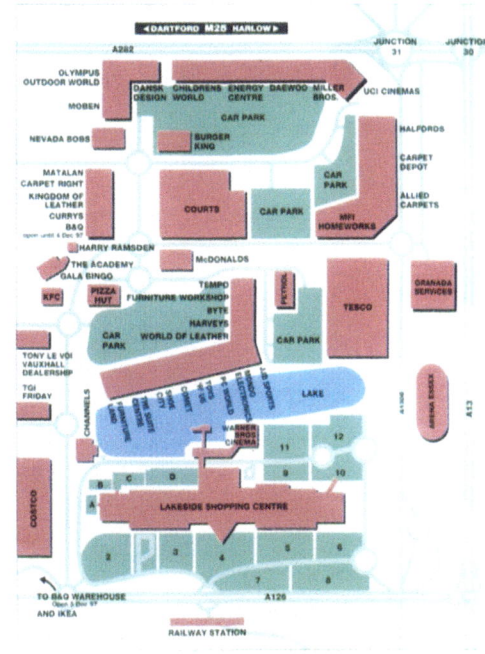
Lakeside Shopping Centre and Retail Park

**Web TV**

WebTV works with the TV
you already own.

It plugs into the standard telephone line
you already have.

And you can use it from the best
seat in the house.

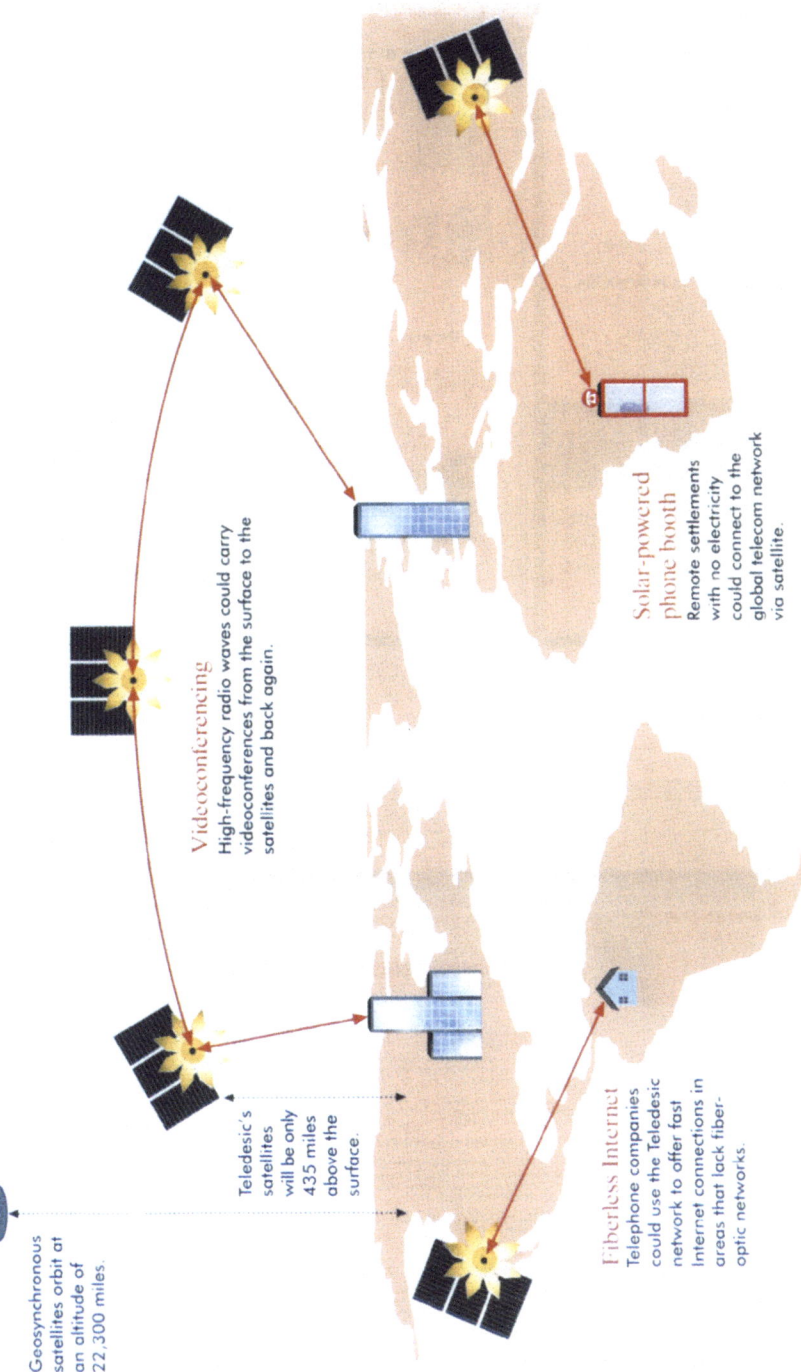

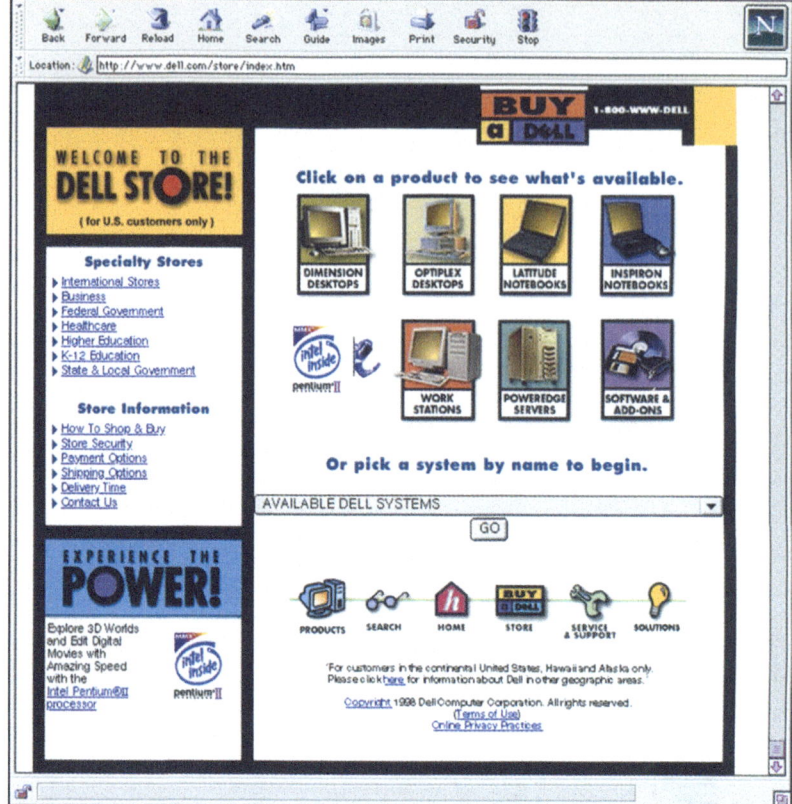

Ordering from Dell over the Internet

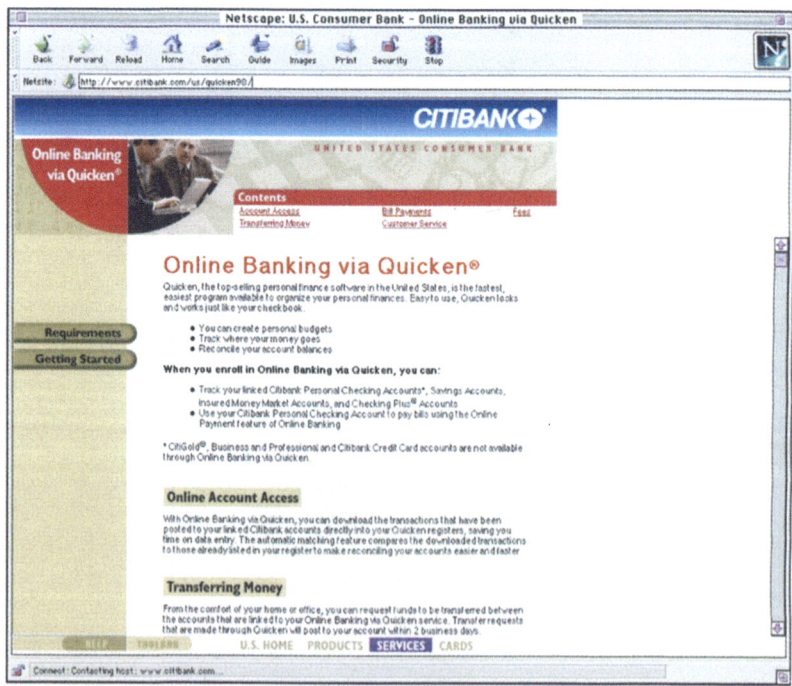

On-line Banking via Quicken at Citibank

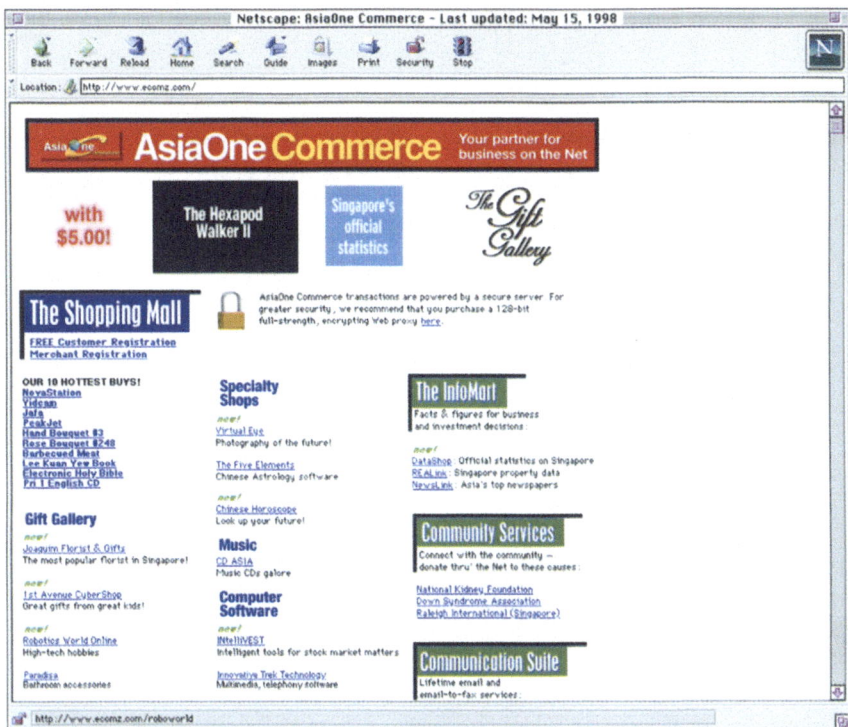

AsiaOne Commerce

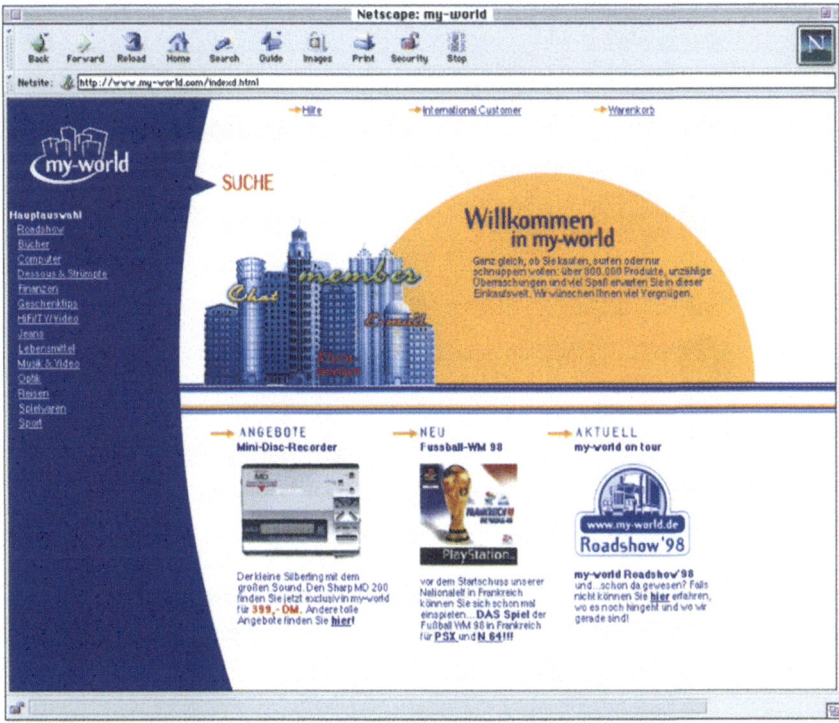

Karstadt's My-World

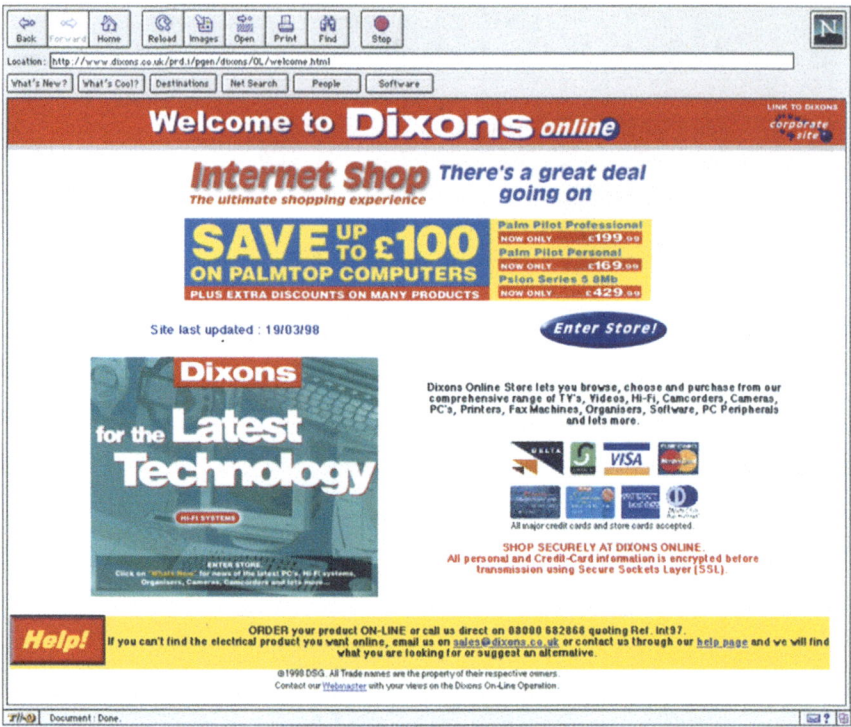

Dixon's website

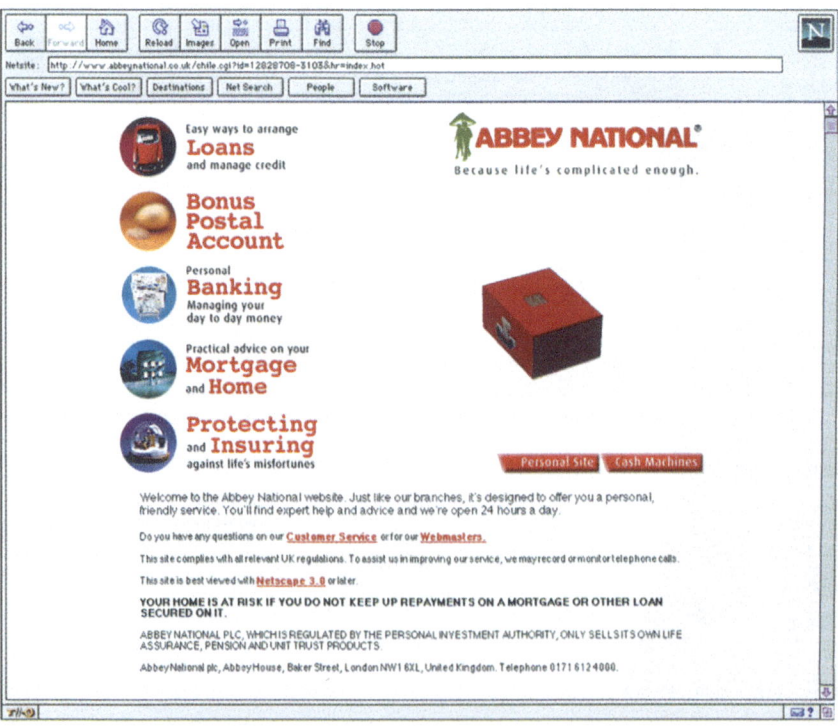

Abbey National website

# 8 The Store of the Future

What will the store of the future actually look like? Before illustrating the 'Best of both' option referred to in Chapter 7, we should briefly examine some of the new retail technologies that are being developed and which enable the future store to take shape.

**In-store technologies**

Stores are now becoming intelligent. It's not just about EPOS and barcoding, the whole store is becoming smart. New technologies like wireless networks, multimedia kiosks and the Internet can be integrated into the store operation. They transform what information can be communicated in-store and how quickly it is transmitted. They can link every employee not just a few sitting in the administration office. They can also, most importantly, involve each and every consumer. The whole store can be brought alive. Consumers can be electronically connected. They can be networked into store information lines and into employees for assistance as they walk around. They can do their own bar-code scanning, pay when they want, avoid waiting in lines where others tell them. They can be empowered, made to feel more in control and given the means to extract the maximum value from the shopping experience.

To facilitate this, there are two particular in-store technology innovations coming through. Each needs to be described to understand the impact they can have on future store layout and design and the services retailers can start introducing to retain and attract customers:

- Wireless networks and hand-held terminals
- Touch screens.

*Wireless networks and hand-held terminals*

If physical cabling is no longer a constraint then retailers are immediately liberated from a fixed store layout. While retailers have over the past ten years invested heavily in underfloor cabled EPOS systems through the store, there is no doubt that any major retailer investing in a new site would now go for a more flexible wireless environment.

There are cost benefits but more importantly significant customer service advantages as well. Sales staff can be equipped with portable PCs linked to the store server. They can deal with individual queries, check on stock availability, identify any order lead times and arrange for immediate replenishment from the warehouse. In some stores, for example, couples about to get married are invited to walk the store with a store assistant, checking what's in stock and compiling their wedding list on the portable PC. No need to wait in line at a counter or go back to the old fifth floor 'customer service centre' to be dealt with.

These wireless systems are also known as RF (radio frequency) and they can be seen increasingly in stores with sales staff wearing headset microphones attached to belt-mounted RF terminals. By 1998 most of the top US retailers had RF systems in some form and while the initial installations were among back-office staff linking warehouse stock-picking, it's now being seen with more prominence in the store front and involving the customer.

'Putting the consumer in control' is an increasingly achievable and powerful new market proposition. For example, customers themselves can be given their own, simple hand-held self-scanning devices which they use to scan the bar codes of the products they choose. The current scheme involves them handing over the scanner on the way out to a checkout assistant who then prints off the list and a receipt for the customer to pay. There is no need to unload the shopping basket. A further evolution of this technology is to make an exit channel which automatically reads a radio tag embedded in the product's packaging, computes the cost of the items purchased and prints out the bill – all in about 30 seconds.

Albert Heijn, part of the Ahold supermarket chain headquartered in the Netherlands, has been one of the pioneers of hand-held self-scanners and many of its stores now offer customers this facility. UK supermarkets like Safeway and Tesco have been quick to trial these devices and say they plan to make them more widely available once screen display and image definition improves.

*Touch screens*

Touch-screen technology is making a significant contribution to the wider consumer acceptance and usage of computers for the purposes of gathering information and shopping. ATMs have already become an accepted way of carrying out basic banking transactions even to the point where the bank branch is, in some instances, being retained but the service staff completely removed. All that's left is a bank of touch-screen machines dispensing cash, taking deposits and enabling customers to transfer money between accounts. For personal advice pick up the white courtesy telephone and speak to a member of staff, possibly located a hundred miles away in a call centre (see also Appendix 2).

Touch-screen formats are fast and can respond to consumers who want to carry out the transaction as quickly as possible. One US retailer, Frank's Nursery, found that touch screens reduced typical transaction times by up to 50% while retaining 100% information accuracy. Touch screens are also attractive because they easily optimise eye–hand co-ordination and can provide a sensory experience to complement what is normally just read on the screen. More sophisticated versions can combine sight and touch with sound as well, beginning to provide a more 3-D virtual reality experience.

This technology is now finding its way into the retail environment in what are called 'multimedia kiosks'. These are often free-standing sites, sometimes enclosed, where people can go to browse through goods displayed on the screen at their leisure and in some degree of privacy and comfort. Banks like Barclays and car manufacturers like Daewoo are but two groups who have persuaded supermarket retailers to provide them with this kind of space. Most kiosks give access to material running on CD-Rom but some are also now linked to a company's Internet site enabling consumers to gather more personalised information and place orders. Who needs a bank branch or a car dealer's showroom in the local town if the information and order facility can be provided virtually from a broader-based, one-stop retail environment as found in a large supermarket or still larger shopping centre? As supermarket retailers specifically face their own disintermediation challenges they are ironically facilitating the physical demise of banking and other retail outlets. By providing virtual space for others they are hoping it will prolong their own physical lives!

Touch-screen trials are pushing out the current electronic commerce boundaries. One trial led by Cornell University is called CU See Me. It provides on-line video broadcasts, accessed via a touch screen, to take place between users to give live product demonstrations. Another example comes from British Telecom who have been exploring public touch screen kiosks as the 'new revolution telephone booth'. There have been mixed results so far but the idea is to connect users to travel information, enabling them to book flights or order flowers and send gifts around the world. Eventually the plan would be to provide full Internet access. These booths could be located in the street, in shopping areas, inside supermarkets, in hotels, colleges, at motorway service centres. BT plans to trial more than 10,000 terminals in the UK over the period 1997–2000.

\* \* \*

These new in-store technologies will be an important aid to retailers in their fight to stay physical and prevent all their custom rushing to a remote virtual shopping environment. It will help them keep their stores fresh and innovative providing new experiences. It will give opportunities to reconfigure store lay out, do away with the rather cumbersome and old-fashioned lines of check-outs and make the in-store environment and service more user-friendly and attractive. They can provide other services in kiosks, put touch screens in the store coffee shops and respond immediately via RF to enquiries about stock or requests for more information. They can embrace the electronic world but do so from their real estate. They can have all the most sophisticated electronic connections that the consumer could ever want, and all under one roof. If they respond positively enough down this path they would certainly have a very strong chance of not just surviving the electronic revolution but coming out the other side all the stronger in their customer franchise.

This potential future world is described in the next few pages showing one particular view as to what an integrated physical plus electronic shopping environment might look like. It's an end-1990s view of a possible future, sketching out what could be achieved as technology advances and matures. It's a vision for a retail group that has thought through its options and made a commitment to embrace electronic shopping but from its physical estate. There is no doubt, though, that as the next decade unfolds the boundaries of what is possible will extend even further, making today's dreams even more realistic and achievable.

### 'Best of Both' illustration

As we drive into this 'store of the future' it already looks and feels very different. In this 'Best of Both' option (also see option 9 in the previous chapter), the goal is to retain the essential integrity of the supermarket as distributor and product showcase while embracing the electronic age to provide a more exciting yet comfortable shopping experience. Let's examine a few snapshots:

**Frame 1:** The parking lot still contains plenty of room to park the car but there are now a number of customer service options

- Self-Parking
- Valet Parking
- Family Parking – with special assistance for young children
- a route for 'Pick Up & Collection' to drive straight round and pick up orders made electronically.

Walk into the store. There's a substantial revision to what services are available and what products are actually merchandised in front of store. There's an integration of electronic shopping needs with those of the more traditional shopper. The aim is to make shopping fun and enjoyable and still retain the necessary functional elements. Shoppers can wander the store or go straight to the coffee shop. They can order electronically at multiple points in the store or they can pick up a cart or basket and fill up with special foods, new products and the evening meal.

The store is divided into a number of sections and in Figure 8.1 there's an open plan of how it might be laid out:

- Reception Area
- 'Home Delivery Service Shop'
- Children's Drop-in Centre
- Local Information Area
- Services Arcade
- Ready Meals
- Fresh Foods and Healthy Eating
- 'Electronic Coffee Shop'
- New Products Centre
- Picking Warehouse and Administration.

**Frame 2:** Reception Area – the store now offers a range of services and activities so as we walk in there's a hotel-style reception area. There's a general enquiry counter/customer service desk helping us find our way around because it's still new and different and we're still learning what this new store has to offer. There's a lobby area where we can meet friends, wait to pick up the kids or otherwise relax.

**Frame 3:** Home-Delivery Shop – just off the Reception area is the Electronic Shopping service counter. It deals with general

**Figure 8.1** Plan of the Store of the Future

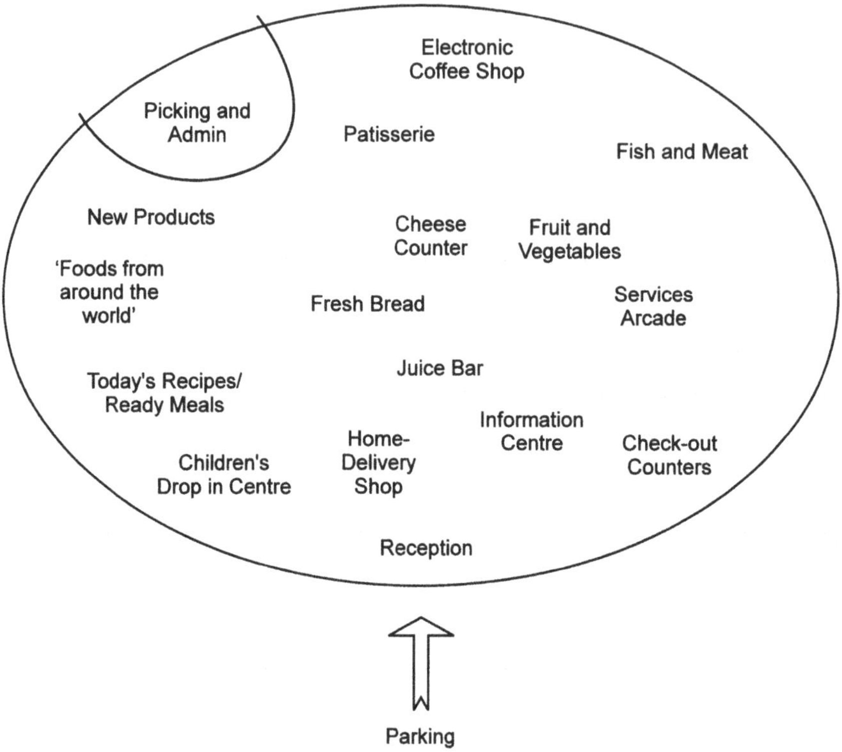

```
                    Electronic
                    Coffee Shop
      Picking and
      Admin         Patisserie
                                              Fish and Meat
      New Products     Cheese      Fruit and
                       Counter     Vegetables
      'Foods from
      around the                              Services
      world'        Fresh Bread               Arcade

                         Juice Bar
      Today's Recipes/
      Ready Meals
                              Information
                  Home-       Centre        Check-out
      Children's  Delivery                  Counters
      Drop in Centre  Shop

                    Reception
                        ⇑
                     Parking
```

queries and provides information on the home delivery operation. It will set up appointments with in-store representatives for customers to come in and set up their standard shopping order for automatic weekly delivery. Those who want to do it themselves can go to special multimedia booths where they can log on to the store's electronic shopping list.

**Frame 4:** Children's Drop-in Centre – this is where the kids can be left under supervised care. It can function as the local community drop-in centre where children can be left all morning or all afternoon while parents are off doing other things. It would have trained staff in a secure environment with learning, education and games facilities. It will help establish the store as a focal point for the community, it can help extend its role from supermarket to one-stop service centre.

*The Store of the Future*  127

**Frames 5, 6:**  Local information area and Services Arcade – before walking into the areas where the products are displayed, two other service areas are drawn to our attention. There is a city-county-wide information area staffed by the local residents' associations providing details on local events, school concerts, market days, etc. And beyond the information desk is the Services Arcade. There's a bank of ATMs, a pharmacy, a travel shop offering detailed advice on holiday destinations, also general Internet services kiosks where we can access the whole world wide web and carry out any other shopping or information needs we have.

**Frame 7:**  Ready meals – recognising consumers' busy life-styles and reduced time for preparing meals at home, there's a ready meals section, New York deli-style. There's a chef providing demonstrations of 'recipes of the day', the opportunity to sample some of the prepared foods and a special counter where all the ingredients for the recipes can be obtained.

Just beyond the Ready Meals area we might find a 'Foods from around the World' counter featuring produce and speciality dishes with regional flavours such as Italian, Chinese or India. Such areas will be typical of the new store. Instead of going to the frozen foods section to buy the pizza and french fries, it will be merchandised and brought together in more colourful ethnic concepts and ready to 'take away'!

**Frame 8:**  Fresh Foods and Healthy Eating – a large part of the store will be taken up with items that don't come in packages and which more naturally appeal to consumers' instincts of touch, taste and smell. There will be fresh bread counters, cheese sections. There may be a patisserie. There may be fresh meat and fish. Each store will find its own selection which best meets local tastes and interests. Some may add an organics section while others may have 'taste and try' counters where people can test out new things.

**Frame 9:**  New Products Centre – manufacturers will remain especially keen to bring their new products, flavours and variants to consumers' attention. We might find a special section we can walk around with such new products on display and manufacturers 'renting space' from the store for a day to demonstrate their new goods, give away coupons and samples and help establish their new items in the consumer consciousness.

**Frame 10:** Electronic Coffee Shop – this will be more of a lounge/restaurant than the old-fashioned-fast food facility. It will be a place to relax, meet friends as well as get breakfast, lunch or a snack.

What will be special about this coffee shop is that both in the lounge area and in the restaurant there will be tables with computer screens displaying the full range of the store's products. People could order their goods electronically seated comfortably at their table while waiting to be served. They could then either arrange to collect those goods on the way out or have them delivered to their home later in the day. If they already have their standing weekly order set up they could simply enter their password codes and send the order through with a couple of taps on the touch screen. It's shopping while relaxing.

**Frame 11:** The back of the store is now converted into a large picking area. Many of the items previously laid out in front of the store on shelves are now in boxes in picking slots. A team of staff have responsibility for picking up the orders electronically, packing them and moving them to the Collection Point. Depending on the amount of space available, the picking area might be located on an adjacent site.

\* \* \*

Will this store of the future, or something like it, become the reality? Will retailers restructure their physical estate so it does embrace the new technologies? Will they 'wire' consumers up, let them do their own ordering at their leisure, provide instant payment facilities instead of long check-out lines, let the customer pick up their packed order on the way out or have it delivered home later the same day? It will require a retraining of staff. Technical capabilities will have to be upgraded. Communication skills and interactions with customers will need to become even more sophisticated. Front line employees will have to learn the meaning of hassle-free customer service!

Leading retailers are well placed to make these bold moves. They have the financial position, the market clout and as we have discussed a history of making step changes in supply. Timing will be critical, the technology must be user-friendly, fulfilment procedures must be in place, the store must be able to easily deliver what it promises. It is certainly achievable. Experimentation down this path is bound to take place. But, ultimately consumer demand plus competitive pressures will surely force the pace and also the ambition of the services and store facilities retailers set out to provide.

# 9 Rapidly Improving Technology Meets Growing Consumer Demand

If retailers don't take the initiative, then consumers will. The 1990s has seen the rise of the 'prosumer' – the more experienced and self-reliant consumer who is self-confident and assertive about what they need, is demanding in terms of service and information and is prepared to proactively search out for value and for suppliers who will deliver the services that are required.

As consumers reach out to take more control of their environment so technology will become their key enabler. The growing spread of the Internet will help consumers shop around to review competing products, sophisticated search engines will compare prices and point out where the best value will be. Choice will range across the whole world and there will be no constraints from limited shop opening hours or drive times and catchment areas.

This combination of more sophisticated consumer demand plus rapidly improving technology are the key forces driving the changes in shopping habits and the arrival of the electronic revolution in retail distribution (Figure 9.1).

These forces have developed an inexorable momentum. Consumers are better educated, have higher expectations and are under increasing stress and time pressure in their lives. They know what's possible and will go after it relentlessly. On the technology side, there are now many organisations across many industry sectors in Computers, Telecommunications, Electronics, Semi-conductors, Software, Publishing and Entertainment who are investing aggressively and are determined to turn electronic commerce into a multi- billion-dollar revenue and profit opportunity. It's a coming together of changing demand with a supply side that's on the move. It's creating the revolution in the world of commerce and this chapter looks at how each will materially drive the electronic shopping market.

**Figure 9.1** The Consumer and Technology

On the demand side, we're looking to be more precise as to how many consumers will be imbued with the desire to buy things electronically. Aren't consumers happy enough with today's shopping arrangements? Where there is dissatisfaction, how widespread is it? There's talk of c.15–20% of consumers – a hard core – who are already impatient for the technology and infrastructure to buy electronically, how ready are they to shop in this way? What's the impact of a 15–20% drop in store traffic on store profitability? Is it going to stay at the 15–20% level?

On the supply side, we're looking at the state of the current technology for electronic commerce. How is it expected to change and what are likely to be the most popular and successful electronic connections? There's talk that current copper telephone wires are limited in their capacity and the cost of replacing them too much for one corporation or consortium. But will that be necessary? Is there technology which can already see past these roadblocks and which will bring data and images at high speed and resolution onto our PC or TV screens and into our homes? How soon will it be available?

**Changing Demand**

*USA Today* in a recent article announced the arrival of the 'I want it now' society – an environment where consumers are becoming increasingly impatient and intolerant, where they know what is possible and are unwilling to accept anything less than the best. The article was commenting on the 1997 strike at UPS (United Parcel Services) in the USA and the delays that were caused in getting parcels and other mail quickly from one place to the next. One of the reasons for the strike ironically was that UPS employees could not keep up with the pace of demand and that consumer insistence on immediate delivery was exerting too much pressure on them. The strike and the outcry that was caused by the delays was all part of a wider syndrome, *USA Today* suggested, that was sweeping America. Consumers were increasingly demanding: 'I need it now or just don't bother!'

> 'In a society in love with the immediacy of the Internet and the speed of the fax machine, overnight delivery services have taken on an overblown importance. We are seeing the emergence of the "overnight society" that is increasing expectations but at the same time adding to the tension and stress in our lives. Technology has been the catalyst that has created this situation. Where companies used to have a week to come up with a big idea now it is expected within an hour over the fax or through the PC.'
>
> *(USA Today)*

As the end of the twentieth century nears so we're moving closer to an environment demanding instant gratification. It's put increasing pressure on all supplier service levels and many have been struggling to keep up with the pace, utilise the improving technologies and meet the underlying consumer needs and demand. As a result consumers are often left frustrated and disappointed with late deliveries, long waits, broken promises, insufficient information and inadequate response to queries and complaints.

It's a situation that leaves a large gap in the market place – a widespread consumer need that still is not being fulfilled. We're just completing the 'customer service' 1990s, but it seems consistent customer satisfaction is still a long way off in many different industry sectors. No wonder then consumer interest is so high in electronic commerce. Any proposition that looks as though it *can* respond to changing consumer lifestyles and *can* deliver quickly and effectively by exploiting new technologies is likely to have a high take-up rate.

*Still a large customer service gap*

For retailers the gap between promise and fulfilment remains too high for comfort. They are leaving themselves vulnerable to new, more customer-effective propositions. The 1998 American Society for Quality annual review of Customer Satisfaction, reported in *Fortune* magazine, shows that only one retail chain in the USA made the top 50 companies for consumer satisfaction! That was Nordstrom. But even Nordstrom couldn't make the top ten (Table 9.1).

**Table 9.1** 1998 Satisfaction Index

*Top 10*
1. Mercedes Benz
2. H. J. Heinz food processing
3. Colgate Palmolive
4. H. J. Heinz pet foods
5. Mars
6. Maytag
7. Quaker Oats
8. Cadillac
9. Hershey Foods
10. Coca-Cola

26. Nordstrom

*Source*: NQRC, University of Michigan.

This same survey also calculated average satisfaction scores for industry sectors at large. The retailing community did not come out especially well, not as bad as the IRS (Internal Revenue Service) but still some significant way off the leading pace (Table 9.2). The survey concluded that the service sector generally had the least satisfied customers – mostly because 'down-sizing had reduced the number of front-line workers and made them stressed out'. In contrast consumers were generally 'happier with manufacturers' and certainly many of the top companies in this report are consumer goods manufacturers who are

**Table 9.2** Industry Sector: Customer Satisfaction Scores

1. Beverages
2. Pet foods
3. Personal care
4. Food processing
5. Parcel delivery
6. Household appliances
7. Consumer electronics
8. Automobiles, trucks
9. Gasoline
10. Tobacco
11. Apparel
12. Insurance

17. Department and discount stores
18. Supermarkets

32. Internal Revenue Service

*Source*: NQRC, University of Michigan.

## Rapidly Improving Technology

clearly seen as fulfilling their end of the supply better than their distributors.

According to a study by Professor Berry at Texas A&M University, 89% of American adults are still 'dissatisfied shoppers'. They can't get what they want when they want it. Many companies, Professor Berry found, just don't do enough to assess and understand what their customers really think, their consumer research is relatively superficial, they struggle to build loyalty and just don't develop the insight and appreciation that would help them retain and build their customer base.

In fact the Technicians' Assistance Research Program (TARP) based in Washington found that consumer complaint was substantially higher than most surveys recorded. In their research they found typically only 4% of dissatisfied customers actually took the time to formally complain – many just changed supplier. For one company they found for every one active voice of dissatisfaction there were in fact 26 more that were never filed or communicated!

*Retailers and others must address the service problem as a prerequisite for succeeding in e-commerce*

With consumers dissatisfied but more demanding, it is inevitable that the promise of electronic shopping will be seen as especially attractive. That is, if it delivers. If it does it cuts out a large part of the frustration. It eliminates the car journey, the car parking, the crowds, the trolleys, the aisles, the long waits, bored unhelpful staff, inadequate facilities. All this can be avoided while consumers sit comfortably in offices or den searching out at their own pace and leisure what they want.

Yet the various sources of consumer dissatisfaction are *not* outside the retail chains' control. There are areas that retailers could turn to and substantially improve upon. They could still make enormous strides in better understanding their consumers, even within the existing physical environment, isolating the problem areas and fixing them. Right now most retailers still appear to pay lipservice to customer service and it is clear from the (dis)satisfaction studies that few have got it right. Their priorities seem more turned to aggressive new store-opening programs in domestic or overseas markets (see Appendix 1, p. 245). They are more focused on new space than getting existing space to deliver a totally customer satisfying experience. But as the electronic revolution gathers pace so new space programs will have to come under increasing reappraisal. With that perhaps at last retailers will turn to their existing proposition and begin to pour their efforts into getting that right and maximising its potential so it can stand up and offer such a compelling proposition that it can resist even the finest dedicated electronic service provider.

*Significant numbers of consumers see ES as a better alternative*

The retailers' task will not be easy. Already consumers are showing high levels of interest in the Internet and other forms of virtual shopping and these interest levels do appear to be at the critical mass 15–20% level, even before the technology and infrastructure has been put effectively in place. Research referred to in Chapter 5 by A.C. Nielsen and others has identified how different groups of consumers feel about shopping today, their readiness to explore new things and their desire and need for convenience.

Taking and adapting their research for an electronic environment they found (Figure 9.2) that 17% of the population at large could be categorised as 'Frenzied Copers' – people who just don't have the time to do what they want and are 'frenziedly' interested in exploring time-saving propositions. Another 16% were categorised as 'mercenaries' – simply out for what they could get switching to whatever offered them the best

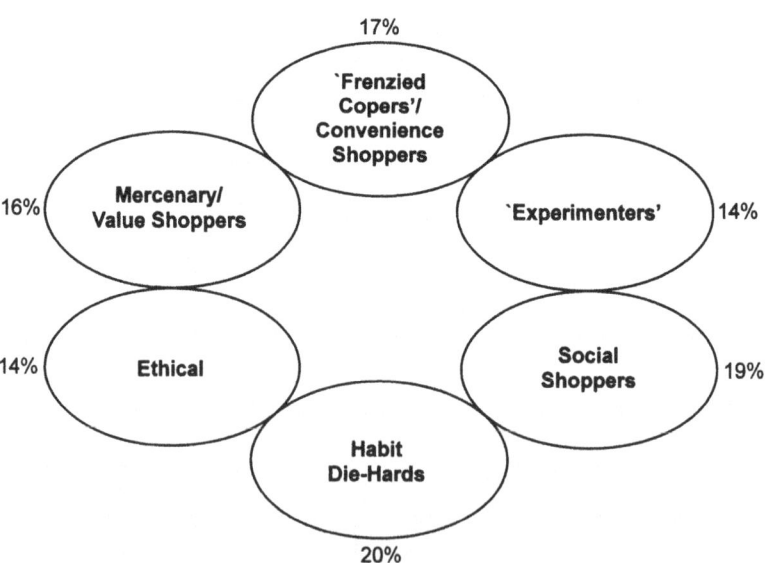

**Figure 9.2** Consumer ES Interest

*Source*: A.C. Nielsen, Mintel, Henley Centre.

value. Another group could be categorised as Experimenters, 14% of the population, who also had little loyalty and were among the first to try new things.

This research suggest that there is a core 17% of the population (the 'Frenzied Copers') who will respond without hesitation to electronic shopping in all its forms. Hot on their heels will be sizeable other groups especially 'Experimenters' and 'Mercenaries', who make up another 30% of the population! Does this not suggest there is really a substantial body of people waiting for the revolution to come? Certainly not all will be immediately interested. There is at the same time a large group who in contrast appear relatively comfortable and contented with existing distribution patterns or are 'stuck in their ways'. But like any major social changes, once they reach critical mass they do build an inexorable momentum that can carry even the reluctant before it.

The A.C. Nielsen research also asked consumers the blunt question: 'Would you shop electronically?' They used grocery shopping as their example. The responses were in line with their original categorisations. 19% of all households said they would (Figure 9.3).

This Nielsen research is not unrepresentative. It's not an isolated view of the market place. Another well-respected consumer research group – Mintel – found in a 1996 survey that 31% of shoppers did not like visiting the stores to get their shopping (Table 9.3). Such shoppers classed themselves as 'reluctant' or 'obstinate'. They saw no social reason or benefits in visiting the supermarket or other stores. They appear especially interested in alternative means of getting what they need.

**Figure 9.3** Readiness to Shop Electronically

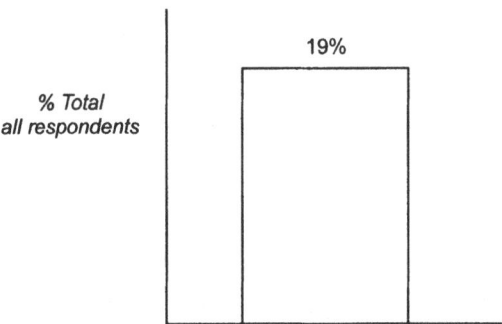

*Source*: A.C. Nielsen.

**Table 9.3** Levels of Shopping Enjoyment %

| | |
|---|---|
| – Enjoy shopping and go frequently | 15 |
| – Enjoy shopping and go occasionally | 19 |
| – Shopping can be enjoyable but only for something I need | 35 |
| – Shopping is not enjoyable and go only when have to ('reluctant shoppers') | 18 |
| – Hate going shopping ('obstinate shoppers') | 13 |

*Source*: Mintel/BMRB.

*Growing 'kiddie power' fuelling ES interest*

While these surveys were all taken among adults, as was pointed out earlier there is also a growing generation of children for whom the computer is as important as the television, who are comfortable in using it, want to use it and readily embrace it as part of their everyday lives. For this generation, while they remain teenagers going out shopping is principally a social occasion that they look to and enjoy. But as they reach adulthood, take on jobs and families and begin to experience the time pressures of everyday living, won't their interest in internet shopping and its advantages be even more highly developed? Won't polls among this group in ten years' time show even higher interest and acceptance levels for buying and interacting electronically? Won't this new adult generation provide a further surge of demand for home shopping, pushing it to critical mass volume levels among the population at large?

In 1998, forecasters estimated there were c.5 million teenagers surfing the Net. According to Jupiter Communications that figure will grow to 90 million by the year 2000. Clinton in the USA, Blair in the UK, Jospin in France and Kohl in Germany have all publicly stated their aims to get all their schools wired up to the Internet by the end of the century. The mass middle classes now increasingly feel they should have a home PC and sign up with American On-Line or one of the other access providers to help their kids stay in touch with the new technologies and access the more sophisticated education and information material now widely available.

Suppliers, too, are seeing the emergence of 'Kiddie power' on the Internet. They are investing both to meet and to fuel demand. As *Business Week* pointed out 'the kiddie market itself will be a cyber gold mine in a few years time'. Media and web giants are scrambling to offer new kid-friendly sites. Companies like Disney have been the latest to bet that 'Net kids' is where the money is and Disney has brought the phenomenon into the spotlight with its Daily Blast site featuring games, stories, comic strips, current news and other services to subscribe to. As they grow up net-literate, won't many surely graduate to doing their shopping in this same way? A recent survey by the Henley Centre suggested 80% of 16–24-year-olds expected to do their banking and food shopping on-line in the future. Two-thirds of all under-21-year-olds still in education already claim to be 'regular Internet users'.

*Whole communities are going electronic*

We're already seeing whole cities and towns embarking on a program to get everyone in their community wired up and inter-connected. We referred earlier to experiments in Palo Alto, in Finland and in the streets of London. But these are just a few examples among many others stretching from Silicon Valley to Malaysia. As the *Wall Street Journal* commented:

> 'many residents in even the remotest towns react incredulously at first to the prospect of being part of the internet community but these erstwhile computer illiterates quickly learn to log-on to check children's school grades, housewives go on-line to compare prices at local stores and order up things for the home . . . even the town's mayor can be found checking a chatline to hear the latest grumblings about potholes and burned-out street lamps!'

*Just a 15% switch can make the difference*

It's all adding up to a major challenge for companies in the decade ahead. Few organisations can take the ES Test and conclude that these changes won't in some way affect their business. And we have to remember that it doesn't require the whole population to change their shopping habits for the impact to be significant and far-reaching. For many retail chains it will only require a reduction in store traffic of c.15% for their stores to be plunged below break-even and into a loss-making situation (Table 9.4). Retail net margins are generally just too thin to sustain substantial shifts in the way consumers buy.

15% of US retail sales would amount to c.$300 billion of commercial activity potentially shifting away from existing physical distribution. In

**Table 9.4** Impact of 15% of Consumers Buying Electronically on Store Profitability

|  | Current |  | 15% reduction |
| --- | --- | --- | --- |
| Store sales |  | 100 | 85 |
| % cost of goods | 75% |  | 76.5%* |
| Actual cost of goods |  | 75 | 65 |
| Gross margin |  | 25 | 20 |
| Fixed costs |  | 20 | 20 |
| Net operating margin |  | 5 | 0 |

Note: *Reduced manufacturer discounts as sales decline.

the UK the figure would be c. £30 billion. This is a substantial slice of retailers' current business. It may be a big jump from Internet shopping levels today but the consumer interest for this is unquestionable given the consistent consumer surveys on the subject. Indeed another research group, Verdict, polled UK consumers and found that around 15% were so determined to access Internet shopping when it became available that they would be prepared to pay a premium for its services. That response was especially strong among ABC1 consumers who have already been at the forefront of bringing PCs and the Internet into their homes.

If the demand is firmly in place, then the key question is: can the technology deliver? What are the range of technology solutions being pursued, how wide a reach will they achieve and are they able to overcome roadblocks to meaningfully enable the first 15–20% of the population – the critical mass – to get on-line and become at last the satisfied consumers they desire to be?

**Investing supply side**

There is an extraordinary level of investment and activity taking place. It's involving companies from many different industry sectors from Aerospace through to land-based cable companies. All are jostling for position looking to drive their technology to the forefront and provide communication into people's homes (Figure 9.4). It's resulting in a wave of alliances as companies get together to combine resources, capital and know-how to accelerate progress and get ahead of competition. For example, in satellite communication heavyweights like Lockheed Martin, Hughes Electronics, GE and Motorola are partnering with others to

*Rapidly Improving Technology* 141

**Figure 9.4** Electronic Convergence

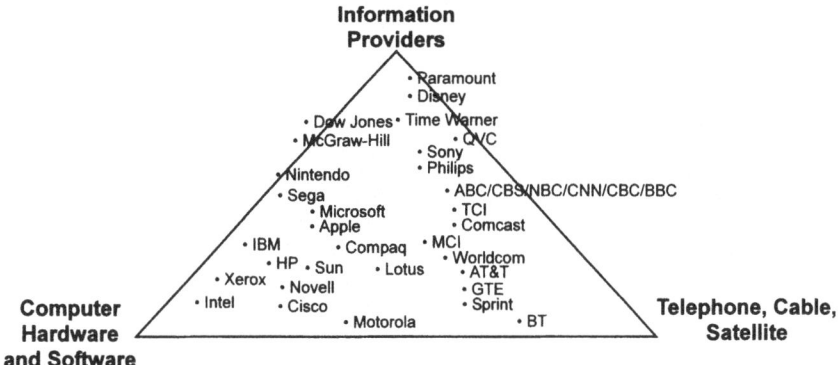

*Source*: New Paradigm Learning Corporation, McGraw-Hill.

bring universal internet communication across the globe. Among telephone companies WorldCom has startled the business community with its audacious but successful acquisition of MCI for $30 billion. This combined group not only now controls 25% of the US long-distance telephone market but more importantly for the future it has cornered 57% of the consumer Internet access market with UUNET the world's largest carrier of Internet traffic, plus MCI's extensive Internet 'backbone' systems network.

There are a host of competing technologies exploiting everything from satellite links to old copper telephone wires and even investigating opportunities using ordinary electricity wires and the basic home-electricity grid system. It's not clear which of these routes will become dominant and to some extent that doesn't matter at this stage. They are all contributing to the development of the 'information market place' and each provides unique advantages as well as having its individual set of investment and infrastructure hurdles to overcome. All these companies in the next few years will bombard consumers with services and opportunities, some claiming their communication links to be superior, others packaging up a complementary set of options providing Internet, telephone, videophone and TV in an easy-priced, readily accessible formula. They are all making rapid progress expanding reach and improving quality of communication. Each route to market will be commercially viable, functioning and affordable in the next five to ten years at a level that can facilitate the practical and serious development of interactive home shopping.

There are a number of mainstream competing routes for providing electronic commerce. Each by itself offers the breakthrough potential to bring ES to the mass population:

- Copper telephone wires
- Fibre optic cable
- Digital Television
- Web TV set-top box
- Satellite
- Wireless cellular phones
- Electricity grid.

Each of these routes is vying for time, space and attention on the home PC or TV or on a screen phone or via an interactive kiosk. It's a complex set of options and opportunities and we need to examine and understand each in turn.

*Copper telephone wires*

Telephony offers the most extensive communications network worldwide. It reaches into almost every household among developed nations, people are familiar with it, it's easy to use and inexpensive. In principle it solves the communication challenge in one, but in practice it's had limitations in the amount of information it can transfer each second – its bandwidth. It has reasonable speeds with text but has proved slow for images, video and sound.

Yet in this global race to win the future communications market, telephone companies have been actively searching for ways to improve and maximise the potential of their established network. One way, which will be examined in the next section, is to replace outdated copper wires with high bandwidth fibre optics. Another way is to learn how to transmit more efficiently and more effectively using the existing copper wire network. Some individual software companies have been especially innovative in this respect:

- *Future Wave Software* has developed an ingenious new way of transmitting data. They compress the time required to send it. The company used to send copies of its images but instead now sends instructions along the wires to recreate its images at the other receiving end. The instructions occupy less bandwidth and so take much less time to transmit.

- Software-company *Progressive Networks* wanted to send video pictures over the Internet but experienced poor-quality results. Typically video is transmitted at 30 frames per second but existing copper-wire telephone lines cannot cope with that. Rather than wait for all its customers to be recabled with high bandwidth fibre, the company developed a new software technology which enabled it to cram more data through the wires and deliver the same picture with similar quality at 10 frames per second rather than 30.

Aside from individual company initiatives there are three particular recent developments, which are being aggressively pursued by the Telcos that are powering up the capabilities of the existing copper-wire networks.

For reference, as we shall start talking about kilobytes, megabytes and gigabytes, some brief definition of these measures of memory size is required. Typically a megabyte can carry c. 5 seconds of television with a good-quality image and up to 60 seconds with a highly compressed image. The highly compressed image may be acceptable where, for example, there is just a talking head and so it can be used for computer-wide windows that are small in size. 1 megabyte can also carry 500 pages of text, 1 minute of good-quality music or 5 minutes of speech. With an ordinary copper telephone line it can take c. 10 minutes to download 1 megabyte, with fibre optic cable it can take 0.1 seconds!

Copper telephone-line capabilities are being dramatically enhanced now by ATM, ADSL and ISDN developments:

- **ATM (asynchronous transfer mode)** is superfast switching technology that boosts capacity on data links. Originally developed by Bell Laboratories for high-speed voice networks, ATM is now being adapted for Internet communications and fast image and data transmission. It's been adopted by the US Defense Department, for example, to relay high-resolution images of Bosnia. The Mayo Clinic in Rochester, USA uses ATM to enable doctors to video conference with other doctors and patients and provide 'telemedicine'. ATM uniquely focuses on switching and routing data at very high speeds known also as 'fast-packet switching'. Its rapid development has been helped because it's been adopted as the world-wide standard and so it allows interoperability of information regardless of the end system or type of information. It's being developed hard by the telecommunications industry because they see it as a way of both exploiting existing copper-wire networks and because ATM allows them to offer not just voice but video and data transmission as well.

- **ADSL (asymmetric digital subscriber lines)** is another new technology, in this case developed by Ameritech. It uses the existing copper-wire networks with an upgrade modem installed in the user's home. It provides a 2 megabit link that allows video and data to be transmitted quickly and at good-quality levels. It significantly improves existing bandwidth capability using digital compression to cram more images and data and transmit them at higher speeds. ADSL increases speed from c. 64 kbps to c. 1.5 to 2 Mbps and latest developments show that might get up to 9Mbps.

  'This represents another boost for Telcos – it's a copper renaissance' the President of US West has commented. A consortium of companies including Intel, Compaq and Microsoft have got behind ADSL technology to agree operating standards and the modem equipment. ADSL is being heavily backed but it does require local telephone exchanges to be rewired and modems installed.

  However there is yet further technology in the pipeline – this time being developed by Rockwell and Nortel – which allows ADSL to work using only existing equipment. Known informally as 'ADSL Lite', a stripped-down version of ADSL is soon to be available for the consumer market. It will involve a much simpler modem connection at the home end which can be retailed easily and plugged in without needing a visit by a specialist technician. While operating at a lower speed than full ADSL, it will still power superfast Internet connections.

- **ISDN (Integrated services digital network)** is already old-fashioned technology – it's been around for nearly five years! It was designed to utilise the existing copper-wire network running between telephone exchanges and the customer's premise. It adds a special component to the end of the telephone line that increases capacity. The component digitises voice or data. ISDN can operate at between 64 and 128 kbps and can be combined to achieve speeds up to 1.5 Mbps (see Table 9.5 for comparisons).

  Because ISDN has been around a little longer than ATM and ADSL its commercial applications have been more developed and the 'special component' links can now be installed in the home or in the office for a relatively small fee (c.$100). ISDN can now also be made available over fibre optic cable ('broadband' ISDN or B-ISDN) and in that environment operating speeds could eventually get up to more than 2 gigabits per second.

  B-ISDN and ATM are now combining to not only maximise bandwidth opportunities but also to ensure the data is routed and switched at fastest speeds.

Table 9.5 Types of Connection Possible to the Internet

| Connection | Speed (000 bits per second) | Time to download a megabyte |
| --- | --- | --- |
| Telephone line | 14–28 | 4–10 minutes |
| ISDN | 64–128 | 1–2 minutes |
| Cable TV modem | 1,400 | 5.5 secs |
| ADSL | 2,000–9,000 | 1–5 secs |
| Advanced Cable TV modem | 40,000 | 0.2 secs |
| Optical fibre | 80,000+ | 0.1 secs |

The goal of these various innovations is to provide broadband capacity that is fully interactive. The aim is not just to send data and images fast and with high quality but allow the consumer to send information back with the same efficiency and satisfaction. The technology is getting there but many of the developments, as of 1998, are barely out of the lab and have not yet been fully commercialised. Even so, it's extraordinary to witness the rate and progress of the innovations that have already taken place in just the few years since the Internet went commercial. If ATM, ADSL and other technologies can achieve the communication goals without the high costs of replacing existing wires with fibre or ringing the earth with satellites, then of course the race to bring the Internet to a mass audience will gather even greater pace and critical mass will dawn even earlier.

*Fibre optic cable*

Fibre optic cable must be distinguished first from coaxial TV cable. TV cable has high penetration already in place, especially in the USA where it reaches c.90% of all households, with c.65% of those signed up and using it. However elsewhere in the world, TV cable's development is much lower reaching an average of only 27% of households in Europe. In the UK some 20% of households have cable laid in the streets outside, but only 8% are connected up to the networks.

TV cable does have higher bandwidth than the telephone line because it can move video but it does not have the same capacity or potential as fibre optic technology. As a result it has to be adapted to enable it to provide fast and satisfactory Internet communication. This is principally achieved by digital compression which can achieve significant improvements in capacity. For example up to 16 video channels could go into the space that had previously carried 1 analog channel. Users, however, need a cable modem or set-top box to enable the decompression and receive and process the text and pictures.

Because TV cable has 40–80 times the bandwidth of the ordinary telephone line and especially in the USA already has a considerable infrastructure, there is a substantial amount of investment and activity looking to exploit that and make it Internet friendly and attractive. Intel, Zenith, General Instruments, Telewest, Nynex and others are all developing cable modems with capacity to transfer internet data at up to 4 Mbps. Current tests show this could be pushed up to 40 Mbps.

As for fibre optic cable, it can send its signals with significantly less signal degradation than coaxial TV cable. It has higher bandwidth capacity and can send more channels into a subscriber's home. A typical fibre optic cable contains 40 strands, each the size of a human hair and is capable of carrying 1.3 mn phone conversations or c.2000 TV channels. A two-hour movie can be sent through a glass fibre in a few seconds. It would take a month over an ordinary phone line.

Up till now fibre optic technology has principally been the domain of the telephone companies, again especially in the USA. They have been weaving high-capacity fibre throughout their networks to facilitate long-distance calls at lower prices. Local telephone companies then used fibre to connect to the long distance carrier's network. As a result in the USA much of the backbone infrastructure is fibre. However local loops into businesses and homes are still the old copper wire with its capacity limitations. Frustratingly, it means that bandwidth available to residential subscribers is constrained by the limited local connections.

Replacing this 'last mile' of copper wire would still require a massive investment. This is true not just in the USA but in other countries many of which don't even have any significant fibre optic backbone infrastructure. For example, a consortium in Malaysia has announced its plans to invest nearly $30 billion in its country (population just 15 million) over the next few years to establish a fast fibre optic information highway.

Not surprisingly therefore telephone companies are turning back to existing infrastructures and exploring what technologies like ATM and ADSL can achieve. Equally cable companies, rather than digging up all the streets with again the massive cost and disruption, are putting significant effort into using digital compression for example to achieve satisfactory Internet communications.

To compare the capacity and potential of each of these developing technologies, see again Table 9.5 for a brief summary of speeds and times. It demonstrates the dramatic gains that will be increasingly available to consumers as these various improvements in capacity become more widely available and accessible.

*Digital television*

While telephone lines and cable look set to be the main foundations of the information superhighway, many companies especially in consumer electricals and in the media are focusing their efforts on getting digital television established and developed for mass consumer appeal. Digital TV broadcasts digital signals by satellite, cable or over the terrestrial transmission network. Digital transmission is attractive because it compresses the amount of bandwidth space the data or image takes up. For example, not every line and frame of a picture is transmitted, only those that change from one frame to the next. Sound and picture quality improve significantly vs analog and many companies have as a result been investigating how they can exploit the technology to bring digital broadcasts and digital shopping to the TV screen rather than through the PC. Their goal is to make surfing the Internet more like watching TV and so make electronic commerce generally and the whole Internet experience immediately more user-friendly and accessible.

Prospects for digital TV have advanced quickly, especially in Europe where there is an open standard known as DVB for digital video broadcasting. This has enabled leading TV broadcasters such as British Sky Broadcasting (BSB) to advance their plans to begin broadcasting digitally in the UK, with a scheduled launch date of winter 1998. In the USA, on the other hand, digital TV has a range of proprietary standards and that is slowing down the market, though Microsoft has recently been leading efforts to change this and establish a common approach.

In the UK, BSB is already engaged in some of its pre-publicity for the digital TV broadcasts. It's in a consortium involving British Telecom, HongKong & Shanghai Bank and Matsushita Electric. It promises to create another 300 channels and bring the cost of owning a channel down to < $60,000. At that price many organisations have enquired about setting up their own TV operation and leading UK retailers such as Sainsbury, HMV, Thomas Cook and Great Universal Stores have already stated their intentions to do that.

This digital operation requires consumers to buy a set-top box as a digital decoder and a keypad that links to both their TV and their telephone line. The telephone link is to make the service interactive, enabling subscribers to the service to respond back to the digital TV broadcaster. Set-top boxes will initially retail in the UK for less than

£300. Advanced decoder boxes will permit access to both digital TV broadcasts as well as digital data transmissions on the Internet. TV broadcasts may be by satellite or existing terrestrial network of TV transmission masts.

Microsoft has not confined its own digital interests to the US market. It plans to offer in the UK free or low-cost interactive digital TV services to users of its next version of its Windows operating system – Windows 98. Users will need a PC that is also equipped with a TV tuner, though at this time few UK PCs in the home yet have that level of sophistication.

Digital TV is expected to develop over the next few years and is forecast to reach c.20% of all television households in Europe by the middle of the next decade through cable and satellite connections. In addition, pressure to switch off analogue transmissions is expected to build, pushing total digital TV penetration to well over 50% by 2010. Like many of these initiatives, success will depend on how effectively the service is marketed, the price of the decoder boxes for the TV and how well it beats rival technologies. One thing in its favour is the involvement of British Sky Broadcasting who have demonstrated their marketing prowess in UK with their single-minded hard-hitting conversion of UK TV audiences from a four-channel to a multi-channel satellite TV environment in just five years.

Many of those initially involved with digital TV saw its potential mainly lying with bringing 'video on demand' into the home. But that service has been trialled with only mixed results. The greater opportunity lies in its ability to bring electronic shopping to a mass TV audience who can pick up the signals at a relatively low on-cost. Customers will be in a position to order their wares using just a keypad which could be little more than a standard remote control service with a click facility. Their signal will then go out through the set-top box up the telephone line.

BSB's plans are to offer the broadest possible range of shopping and other services. They are actively encouraging the involvement of banks and retail groups in an effort to create one of the first mass audience 'electronic shopping centres'. Internet services and games will also be made available. BSB itself forecasts 5 mn customers by 2002. The reach of the TV may prove at least as attractive as rival PC-based initiatives.

*Web TV set-top box*

Web TV (see colour plate) is intended to be a simple Internet connection to the TV screen. It's marketed by the likes of Sony and Philips and utilises the existing TV and phone line providing Internet access through a slimline receiver. Surfing is carried out through an adapted remote control. The main difference from Digital TV right now is that the connection is mainly made over the telephone wire rather than by a TV broadcast. By 2003 the set-top box could alternatively receive signals by satellite or terrestrial broadcast. Eventually the decoder box will be incorporated into the digital television itself.

But over the next few years Web TV represents an immediate user-friendly opportunity to get consumers wired into the Internet. Early investors have included Citicorp and not surprisingly Microsoft hedging its bets with involvement in yet another one of the communication technology options. After slow sales on launch in 1996, Sony and Philips plan to ramp up their own marketing with an estimated $60 million spend in 1998 and a push to lower the price well below the initial $300 ticket.

*Satellite*

Satellite communication promises universal access for the Internet around the globe. It has no need for expensive wire or cable infrastructure. It can reach the remotest island or hilltop. All that's required is to ring the earth with enough satellites to enable this communication through the skies to take place!

Yet this dream is fast becoming reality with as many as 15 different American-based consortia developing plans and looking to invest as much as $40 billion on their systems over the next few years. They will be operating in a newly opened part of the radio spectrum called the 'Ka-band' which purports to offer high bandwidth communication and do that relatively inexpensively between any two or more customers on earth. It promises to transmit data up to 800 times faster than current analog phone systems, for example transmitting an entire 64-page edition of the *Wall Street* Journal in under two minutes vs the hours it would take on existing phone lines!

The leading satellite consortium so far is Teledisc which has the backing of Bill Gates – again – but this time in his personal capacity. Teledisc's plan is to ring the planet with 840 satellites at low altitude. They are in competition with a second consortium led by Motorola and Loral Aerospace but they will use a different number of satellites and different frequencies. Both teams expect to be operational by 2002–2003, but they still have a long way to go – not least of which is launching the satellites!

The pioneer at Teledisc is Craig McCaw, who is almost evangelical in the benefits he foresees:

'... here's a poor village in Guatemala . . . they have electricity, they have TV, they see the wealth of developed nations but they can't afford them. The don't have the communications in place yet they are rich in agriculture and local produce. They could become part of the world market if they could more easily join in the global communications network.'

But McCaw does not only have to manage both the financing and logistics. The other satellite consortia include such heavyweight partnerships as Hughes Electronics, AT&T, GE and Lockheed Martin. They plan, in contrast to Teledisc, to use fewer satellites and instead have a handful of giant geosynchronous satellites thousands of kilometres above the earth rotating in a fixed orbit. Yet they all face perplexing technical problems and will have to watch while other terrestrial-based Internet and communication operations continue to build their own networks and momentum. An extraordinary race between the different technologies is taking place, with huge bets being made that one will ultimately triumph as a dominant mass-volume communicator.

*Wireless cellular phones*

In the James Bond movie, *Tomorrow Never Dies*, Bond swaps his gun for a mobile phone. The special phone not only makes calls but can drive a car, crack open a safe and even take photographs. While it will be a long time till that type of phone can be purchased at the local store, the mobile phone is fast evolving, as articles in journals like *Banking Technology* are pointing out, from just a device to make and receive voice calls. There's now a new generation of small portable electronic hand-held phones emerging which can also plug into the Internet and receive a whole stream of data services.

The term 'mobile commerce' has been coined to describe the delivery of electronic commerce capabilities directly into the hands of the user via wireless technology. It's predicted that the new phone will act as yet another catalyst encouraging consumers to shop electronically and its particular advantages are its easy portability and familiarity.

By 1998 almost a dozen companies had begun trials building off GSM – the global system for mobile communication. It is the European standard for digital cellular telephone systems in operation since 1992 and offering teleservices and wireless data services internationally. These trials include elements of electronic commerce collaborating with banks and retailers to provide information, shopping, share-buying and retail banking at the touch of a few buttons.

One of the trials was set up in 1998 by Nokia, giving 100 users in the Netherlands access to banking and commerce via the Internet using the Nokia 9000 Communicator. It's a portable phone and personal organiser equipped with the software to browse the Internet and send and receive e-mail. The price is £1500 but it's an early prototype and like many such items can be expected to come down rapidly once the technology is fully commercialised for volume sales. Transactions are carried out using a smartcard reader to send and spend electronic cash.

Nokia is not the only heavyweight investigating this medium. AT&T also launched its Pocket Net Phone in early 1998. This sells for less than $300 and is intended to act as a cellphone, e-mail reader and Web browser. The phone can be purchased for free if the customer signs up for the basic services at a flat $30 monthly fee. The Pocket Net Phone uses digital technology though sharing the analog cellular network with voice calls. Early users praise its pocketability but have found the screen and keyboard doesn't yet provide clear enough resolution for typing in messages.

Yet another new screen phone product comes from CIDCO Inc. out of California whose –product the iPhone – was also launched early in 1998. This is a $500 phone that can also surf the Internet but it's larger and has c. 7.4 inch lap-top like screen and mini-keyboard. It rivals the fastest PC modems and was expected to be in wide distribution as 1998 unfolded. Among other interested parties Smith Barney was expected to announce plans to provide iPhones to all its retail brokerage customers. CIDCO executives see their target market as the 50% of US households that don't own PCs, providing an easier route to the web. 'Our vision is to own the kitchen countertop,' claims CIDCO's marketing chief. 'We believe Web Phones have a much better chance in cyberspace than gizmos like Microsoft's Web TV'.

Meantime other companies are seeing the opportunities and jumping into this arena. Philips and Northern Telecom have set up a joint venture and Alcatel is busy in Europe putting together what it calls its 'Screen Phone' which will be a touch-sensitive screen and pop-out keyboard connected to either a fixed or mobile phone.

'Mobile commerce' may be in its infancy at the time of writing this book but as it develops and the equipment and technology mature it will doubtless encourage yet more groups of consumers to try out what it can do. Even the most technophobic may find it hard to resist its simple charms!

*Electricity Grid*

An unexpected alternative technology emerged toward the end of 1997. UK-based Norweb Communications together with Nortel, the Canadian Electronics Group that specialises in telecommunications equipment, announced they had discovered a way to deliver Internet services to the home, tapping into the electricity grid system.

The new technology prevents the electrical current distorting Internet signals and other computer data and has been developed so that electricity companies could offer their customers Internet access at speeds 30 times greater than today's high-speed modem links. Customers would simply have to plug their PCs into their nearest household electricity point to link up to the Internet. The only set-up cost would be c. $250 to put in an appropriate PC card.

Is this the Holy Grail of mass internet communication? It would appear the technology provides immediate Internet access to every household in the land and, at the time of writing, the technology was undergoing a six-month marketing trial ending early in 1999. Norweb and Nortel then believe they will be ready to build it into a 'serious volume business'.

\* \* \*

With so much supply-side investing going on and such a range of technologies being developed there is no question that the latent consumer demand for convenience will be easily met in just a few short years. Even if consumers wanted to, how could they resist? They will be marketed to from all sides by TV broadcasters, telecommunications companies, software houses, computer hardware manufacturers and even by their local utility offering a packaged electricity supply plus internet service. The investment taking place is enormous and the commitment to build the new electronic age growing with each new product advance. The coupling between these new improving technologies and time-pressured, demanding consumers is still in its early days but it looks set for a long marriage and commercial prosperity!

# 10 The World is Changing: Assets to Knowledge

The advent of electronic commerce is heralding a new world order for business. The traditional rules of competition that generations have become familiar with are having less and less relevance. A tide of technological, regulatory, geographic and political change has come about since the early 1990s. Scale can no longer be relied on to provide competitive advantage, many products and services are seen as commodities, new competitors are emerging from surprising places and with dedicated efforts are quickly winning market share, formal borders and barriers are being swept aside, customers have gone truly global in their aspirations, technology now makes possible what once was only dreams and there's a virtual environment fast becoming reality.

The commercial world is changing from an industrial economy based on physical assets, factories, roads and heavy equipment to a new economy built on silicon and computers. A transformation in our lives is taking place that's as revolutionary as the eighteenth-century shift from agriculture to industry. It's bringing with it new dynamics and new drivers for success (Figure 10.1).

This chapter looks to provide some perspective around this fast-changing commercial scene. For a moment let's stand back from the helter skelter of week-by-week ES initiatives and announcements. Instead it's helpful to recognise the historical context and consider the underlying trends and dynamics fuelling these changes. What's the impact for business at large of operating in a digital economy, what does it take to become 'knowledge-based' rather than 'asset-based' and what are the implications? Finally what new economic and financial levers emerge

**Figure 10.1** The Changing Commercial World

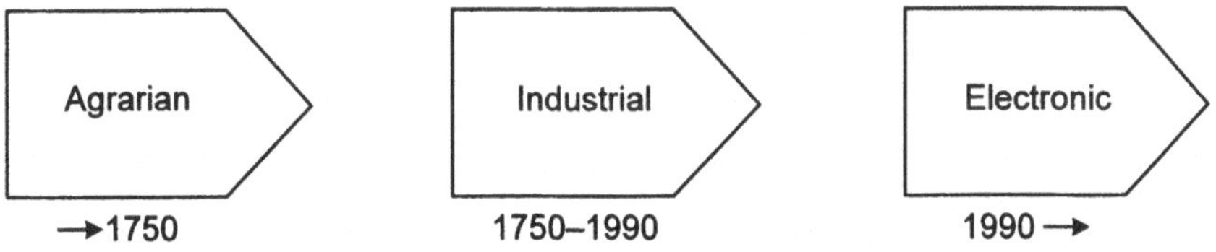

| Agrarian | Industrial | Electronic |
|---|---|---|
| →1750 | 1750–1990 | 1990→ |

with which business must now come to terms and aim to take advantage of?

Electronic technology generally is moving at a remarkable pace with remarkable advances in just a few short years. The major breakthroughs have come from developments in solid-state electronics and the microprocessor. The number of transistors per chip is doubling approximately every eighteen months (Table 10.1). It represent an enormous increase in the power available to drive information technology applications in all dimensions. By end-1997, the world's chip population had risen to 350 bn, including 15 bn microprocessors. That's more than two silicon 'brains' for every person on earth!

There appears no limit to these technical advances. Andy Grove, CEO of Intel, predicts continued progress for at least the next fifteen years. By 2010 he envisions microprocessors with a bn transistors chewing through 100,000 MIPS (mns of instructions per second). That compares with today's fastest Pentium Pro Chip with 5.5 mn transistors and speeds of 'just' 300–400 MIPS.

For retailers these digital developments pose particular challenges. Their industry sector especially is built on physical foundations. It's asset-based. It's all about 'location, location, location' – about land and buildings and physical logistics involving warehouses and fleets of trucks. But such assets become largely redundant in the electronic world. Intangibles like knowledge capture, information management and brand equities become the new core competencies and drivers of value. It's a sea

Table 10.1 March of the Microprocessor, 1971–2010

| Chip | Debut | No. of transistors | Initial MIPS |
|---|---|---|---|
| 4004 | 1971 | 2,300 | 0.06 |
| 8008 | 1972 | 3,500 | 0.06 |
| 8080 | 1974 | 6,000 | 0.6 |
| 8086 | 1978 | 29,000 | 0.3 |
| 8088 | 1979 | 29,000 | 0.3 |
| I286 | 1982 | 134,000 | 0.9 |
| I386 | 1985 | 275,000 | 5 |
| I486 | 1989 | 1.2 million | 20 |
| Pentium | 1993 | 3.1 million | 100 |
| Pentium Pro | 1995 | 5.5 million | 300 |
| 886 (Forecast) | 2000 | 15 million | 1000 |
| 1286 (Forecast) | 2010 | 1 billion | 100,000 |

* MIPS = millions of instructions per second.
Source: Business Week; Intel.

change in the fundamentals of their business. If a new company were starting out today to distribute food, clothing or financial services to the consumer, would it be planning to build expensive real estate? Would it not set its sights very differently? Looking to the future, wouldn't it more likely focus its investment on communication technology providing immediate reach to a total universe of customers at a lower cost of entry? It would surely find it more attractive to develop a virtual rather than physical infrastructure.

The future business model will be built around these very different success factors:

'The traditional factors of production – land, labour and capital have not disappeared. But they have become secondary. They can be obtained and obtained easily provided there is knowledge. Knowledge is a utility, the new means to obtain social and economic results. Knowledge is becoming the only meaningful resource.'

(*Peter Drucker*)

Knowledge has become power, and it is estimated by Drucker that more than 1 trillion dollars per year world-wide is being invested in developing new information and communication technologies, software and hardware to exploit knowledge as a driving source of competitive advantage. Drucker also estimates that one-third of all capital investment in developed countries in the last thirty years has gone into data, information and communications equipment. Nowadays three-quarters of the labour force in most developed countries can be categorised as primarily involved in knowledge work or service. Forty years ago it was less than one-third:

'Leveraged intellect and its prime facilitator, service technology are reshaping not only service industries but also manufacturing. Overall growth patterns, national and regional job structures and the nature of international competition are all being affected.'

(*Peter Drucker*)

The twenty-first century corporation will gradually take on a very different shape. A significant part of its activities will take place in a virtual realm. Physical constraints will be limited. Compaq, for example, can close all its sales offices and instruct its salesforce to work from their homes instead, with computers linked to comprehensive databases. Ford can build its new world car with a virtual team of designers, engineers and others remaining in their different locations scattered across the globe but linked together electronically in a high-tech CAD/CAM video-conferencing virtual environment. Goodyear can check the quality of its raw rubber materials at source and immediately transmit changes in quality or specification directly into centralised flexible automated manufacturing plant. Toyota can outsource even more aggressively and confidently, farming out critical parts of its design and components supply to third parties but with the ability to easily track and check and audit exactly what's going on, on a real-time basis even if the third party is thousands of miles away. Distance will be no object. Corporations need have no boundaries as they reach out to access and integrate allied skills and competencies.

Forward-thinking companies are abandoning old organisation structures and are already aggressively building their virtual world, looking to establish new sources of market opportunity and competitiveness (see more in Chapter 15). But as they evaluate their options, the economic model they consider has fundamental differences. If the value-added levers are virtual rather than physical, are about knowledge rather than assets, then why establish high levels of fixed-asset costs? The virtual environment helps keep costs variable. If fixed assets are required, then they can more readily be outsourced. They can be left with a third party. The company itself can stay flexible and virtual. It can focus, without physical distraction, on leveraging the electronic interactions with its customers. Let fixed assets, mostly in manufacturing and warehousing, be managed by fixed asset specialists. Free up other corporations to establish their value networks (Chapter 6) and find their own key leverage points.

Retailers, for example, who wish to take advantage of the ES potential and gradually transition more of their business down that road will be looking at a very different P&L (profit and loss account) (Table 10.2). Key cost drivers will change and there will be more operating margin flexibility.

A significant area of fixed costs in the Stores is removed. A far greater proportion of the costs is in making each customer interaction more successful. Cost items such as Delivery and Customer Service become significant (though these can be made to be largely variable driven by the number of those customer interactions). Branding becomes a more critical lever. In effect, the brand value is replacing a large part of the asset value. Where previously it was the location that attracted custom now it's the awareness, reputation and emotional values that the seller builds up, communicates and exploits.

## The World is Changing

**Table 10.2** Changes in Retail P&L – illustrative only

| Annual P&L items | Conventional retail store ($000) | As % | Electronic 'retail' operation ($000) | As % |
| --- | --- | --- | --- | --- |
| Revenue | 20,000 | 100 | 20,000 | 100 |
| Gross margin | 6,400 | 32 | 6,400 | 32 |
| Warehouse costs | (600) | 3.0 | (200) | 1.0 |
| Store costs | (4,200) | 21 | (1,502) | 7.5 |
| Delivery | 0 | – | (1,185) | 5.9 |
| Customer service | 0 | – | (200) | 1.0 |
| Brand advertising | (160) | 0.8 | (500) | 2.5 |
| Contribution | 1,440 | 7.2 | 2,813 | 14.1 |
| Overhead | (600) | 3 | (400) | 2.0 |
| EBIT | 840 | 4.2 | 2,413 | 12.1 |

*Source*: CSC Kalchas.

This virtual world will certainly put a premium on brands. It will be intangibles like brands, designs, patents and trademarks that will become the financial bedrock of the future corporation. They will replace the pivotal role of fixed assets on the balance sheet. They will become the capital employed that investors seek to leverage. In a world going increasingly electronic it is surely these items and their balance sheet value that will provide a better guide to the company's market valuation and potential.

US accounting rules still prevent this kind of intangible asset valuation but some commentators are already calculating and quantifying their view as to what US brand assets might be worth if they were translated onto the balance sheet. A 1996 study, for example, by Interbrand and *Financial World* identified how much the top ten US brands would be valued at (Table 10.3).

Some of these brand values are truly massive and even without being incorporated into a balance sheet they offer a compelling view of the value of the company and the power the brands can play. For example Coca-Cola's fixed asset base in 1997 was $4.4 billion, but that bears little relationship to the underlying value of that company: it all resides in the Brand. As former CEO, Goizueta commented:

> 'If all our fixed assets and buildings burned down tomorrow we'd have no trouble rebuilding Coca-Cola based on the strengths of our brands and trademarks alone.'

Table 10.3  Top Ten US Brands

| Brand | Value $m |
|---|---|
| Marlboro | 44,614 |
| Coca-Cola | 43,427 |
| McDonalds | 18,920 |
| IBM | 18,491 |
| Disney | 15,358 |
| Kodak | 13,267 |
| Kellogg's | 11,409 |
| Budweiser | 11,026 |
| Intel | 10,491 |
| Gillette | 10,292 |

UK accounting rules have already moved to recognise these intangibles, acknowledging their value rather than treating it, for example, in an acquisition context as purely goodwill to be written off. While there remains some dispute about the valuation rules, nevertheless a number of UK's leading consumer goods companies such as Grand Metropolitan, Guinness (now both merged in Diageo), Cadbury Schweppes, Reckitt & Colman and United Biscuits have all calculated brand values and incorporated them as part of their net financial worth.

Should companies resist this opportunity to recognise and acknowledge intangible values? Should US accounting rules prevail? Or should the business community rather accept this as one of the shifting realities of the late twentieth century? Are UK accountants for once in fact more prescient than some of their supposedly more commercially-minded US colleagues? Is the current uncertainly about whether to allow brand valuations simply a symptom of an era transitioning between one economic order and the next, waiting for the new rules to be established? Why should Amazon.com, 1-800-FLOWERS and others be penalised for having lower net worth than their physical asset rivals? Why should tangible assets be counted and intangibles overlooked? Why value the custom-pulling power of the real estate and not the same power of the brand equity?

As the electronic age matures so these issues and questions will necessarily be resolved. As companies are forced to put more of their investment into systems, knowledge capture and dissemination so the pressure will mount to have that capitalised value recognised. As companies move out of real estate into virtual economics so they and their investors will be keen to see the net worth of the corporation maintained and hence *all* assets counted.

## The World is Changing

The future world of commerce will be very different from today. Business will develop a totally different view about how to reach customers, how to transact with them and the infrastructure required to do that. The P&L will have a different emphasis, the balance sheet will recognise the currently unrecognisable and investors will develop a different set of tests for measuring the future potential of a corporation. The external market place changes will not only drive the economics but also impact the very shape of the organisation itself, changing the way things are structured, the skills that are required and the processes that need to be developed to interact efficiently with the customer. While the transition is taking place to the new electronic order, the environment for doing business will be more uncertain than it's ever been, employees will need to become even more flexible and ready to embrace change and a company's senior management will have to be ever more adept at planning their way through these changes and demonstrating leadership and clarity of vision in so far as it's possible. It's a changing world and the winners will be the ones who are in the vanguard of changing with it.

# 11 Structural Difficulties with the Internet will be Overcome

On the demand side, significant numbers of consumers are ready and waiting. On the supply side, technology and communication links are rapidly improving. All that is now required is a satisfactory infrastructure that can reassure consumers that their privacy will still be protected, their payment methods secure, that shopping in cyberspace will be as safe as anywhere else. This necessary reassurance is the final piece in the jigsaw. It will be the coming together and maturing of all the three elements – demand, supply and enabling infrastructure – that will trigger the surges of growth that are predicted for electronic shopping. The fusion of consumer interest plus accessible technology is perhaps the most critical. But underpinning that 'marriage' there needs to be a transaction infrastructure that is secure (Figure 11.1).

In this chapter the key security issues are investigated. It is a matter of particular concern to the growth of Internet commerce as that is of course the least mature of the electronic shopping options. The key question is whether that medium especially can be made safe enough to give consumers sufficient confidence to shop. Can a satisfactory environment

**Figure 11.1** Demand, Supply and Enabling Infrastructure

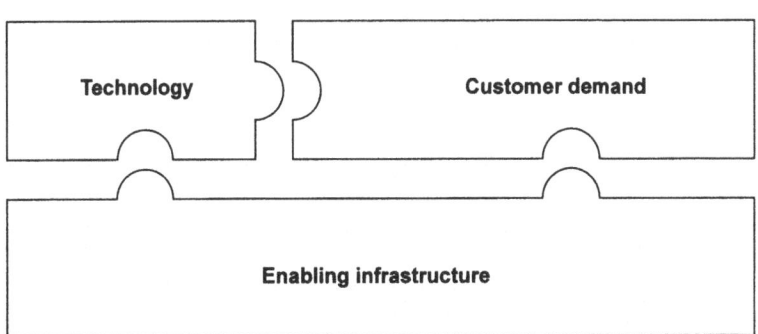

be created? What progress has been made to date, how far away are we till the necessary controls and reassurances are firmly in place?

Some argue that the Internet will never succeed as a commercial medium. They claim that credit card protection will never be totally secure, that messages can be easily hacked into and subverted. But since 1995 there has been extraordinary progress on many fronts. There have been rapid advances in the use of encryption and data protection, for example, and suppliers and banks have worked hard to establish common standards and protocols to better manage and prevent payment or credit card fraud. We will examine this in more detail but the bottom line is that shopping on the Internet looks likely soon to be as safe as any other method of buying goods and services – if not more so.

The security infrastructure issues break down into four key areas and it will be important to examine just how effective the various new technologies and initiatives will be. Which of the many different security systems being developed will be most widely adopted? Will there be common standards that all banks, providers and suppliers can subscribe to? How user-friendly will the security options be and can they actually be developed so they make shopping and payment actually easier and more fun?

The four areas to examine are:

- Personal Privacy
- Message Security
- Payment Security
- General Regulation and Control.

As each of these areas is explored we should bear in mind that these issues are more executional than strategic in nature. In other words they relate specifically to how shopping will get done, they don't by themselves negate the underlying market demand and strategic opportunity. If it's clear that these security hurdles can be overcome then it would appear there is nothing left in the way of realising electronic commerce's full market potential.

**Personal privacy**

Even without the Internet, we already live in an age where information about individuals is kept on any number of databases. When we take out insurance, set up bank accounts, register with a doctor, apply for credit cards, subscribe to monthly information services, join clubs, join air miles or other reward schemes we are listing out information about ourselves. Yes, we can tick the data protection box insisting we want that information kept confidential and not communicated to third parties or sold onto others for direct promotional mailing. Most times when we tick the box we are dealing with a trusted provider whom we can be sure will respect our rights – but we can never be 100% sure. Also we can't be totally safe from determined expert hackers who have set their sights on breaking into information databases to check out or extract information.

Our world is full of information and despite reassurances and regulation we have to make the trade off. We have to decide whether we want the services or products on offer, can vest sufficient trust in the provider to do all that's possible to protect the information we provide and at the same time accept the very small risk that that information might be abused.

Privacy of information on the Internet must be set in this context. It's potentially a medium with no greater or less risk in information management than any other. Furthermore because of the investment attention the Internet is getting, there are mechanisms being put in place to provide additional forms of protection which are over and above the norm. Providers of goods and services are especially keen to address this situation because they are determined to remove any roadblocks in the way of electronic commerce. But we are unlikely ever to arrive at a state of perfect protection either on the Internet or with any other communication/information environment.

Governments have long been concerned about the privacy of the individual and tried both legislative and voluntary routes to try to control how information about people is held and on what basis it can be communicated. As far back as 1980, the OECD (Organisation for Economic Co-operation and Development) established a set of guidelines to assist in the protection of personal data. The principles contained in these guidelines are commonly referred to as the Code of Fair Information Practices and have formed the basis of all privacy legislation world-wide. They include goals of:

'collecting only the information that is needed, inform the person of the purpose of collection, use the information only for the purpose intended, keep the individual informed and give people the opportunity to access the information and get it corrected.'

Building on that 1980 code most western governments have passed some basic privacy legislation. For example, the UK has a charter of data protection rights, in Germany it is unlawful among other things to sell information lists with people's names and addresses without their consent and it's also unlawful to mail people unsolicited material. In the USA in 1986, the Electronic Communications Privacy Act was passed, designed to protect access and misuse of electronic information, reinforcing common law personal data privacy rights protecting individuals. More recently, in 1997, Germany became the first nation to establish a multimedia law specifically aimed at regulating privacy on the Internet, requiring disclosure as to how any collected information will be used and certification that a company's use of personal data does adhere to set standards.

However the world of electronic information has moved on so rapidly that governments are generally struggling to keep up and typically don't have the resources to adequately police how specific companies use and control personal information themselves. Fortunately, though, there are new initiatives coming along which can supplement government attempts at regulations and provide some further reassurance that personal information can be managed effectively and kept private.

One initiative is a kind of self-help device which comes out of responsible people and businesses wanting to act to create the right environment. It's something that can probably only be done easily on the Internet and is an example of how the flexibility of that mechanism can also work to help protect the individuals. What's developed have been self-help pages like the web home page: 'Blacklist of inappropriate advertisers.' That web site lists out all those companies who have offended the voluntary code of fair information practices. It details what they have done and actively encourages consumers not to deal with them. In addition, the US Federal Trade Commission is vigorously encouraging self-regulation. A recent spot check on the top 100 US companies found that 43 of them did display privacy policy notices. But federal regulators want all organisations to voluntarily adopt this sort of code of practice. Otherwise more legislation is threatened and there are currently 32 bills in US Congress all looking to centrally control Internet information and communication.

The second initiative comes with the development of new software that protects the individual from being information-abused. For example 'Net Nanny' catches certain words or phrases such as 'where do you live', 'what is your telephone number' and shuts down the computer, preventing the Internet user from accidentally or mistakenly providing information they subsequently wish they hadn't.

The third and most powerful development has come from the big two makers of software for browsing the web – Netscape and Microsoft. Unusually these two have joined forces specifically to protect the privacy of consumers who visit world-wide web sites. The aim, originally put forward by Netscape in conjunction with two of its software partners, Firefly and Verisign, is for both the big two companies to adopt an open profiling standard that lets users of personal computers manage the information they want disclosed to or withheld from a particular web site. The new technology ensures that any information which might have been automatically exchanged between different web sites is restricted. Only that which the user specifically authorises can be communicated. Industry observers believe the alliances on this technology will make it possible to provide significantly better data protection. 'This is just what was needed out there right now', the Forrester Research group has commented, 'it's going to be the starting point that will transform individual privacy on the net'.

In addition to these developments there are a host of other improvements as we shall see in encryption, use of passwords and methods to protect credit card data and credit payments that are all moving forward with the same intent and the same rapid pace. They will evolve to establish a series of mechanisms that will provide measures of personal reassurance and security on the Internet that will be at least as effective as those found elsewhere.

**Message security**

On top of the protections being put in generally around information and personal privacy must be added specific measures to manage message security – that is, who can access the information or communications being made. There are three particular initiatives – (1) encryption, (2) biometry and (3) firewalls.

*Encryption*

Encryption is simply a system for translating message data into code that cannot be read by anyone without the 'decryption' key. It's an idea that's as old as history and now IBM has established a world-wide standard – the Data Encryption Standard (DES). The software for it is readily available at no cost to anyone with access to the Internet. Browsers indicate that they are in secure encrypted mode by displaying a lock icon, in the lower left corner in Netscape and lower right corner in Microsoft. Early versions of DES were available as early as 1975 and it is now the most well known and widely used cryptosystem in the world. It is especially reassuring that this system is still recognised and has withstood the test of time. More recently it has been enhanced with what is called 'triple DES' which encrypts each message using three different keys in succession. It will basically stop anyone trying to hack in. They would now require not only extraordinary determination but also massive computing power.

Encryption is increasingly used to protect credit card data. It is often linked with additional security measures such as verification procedures to check signatures or passwords that are unique to the consumer. Of course encryption and verification procedures can slow things down – behavioural psychologists estimate that consumers will tolerate a wait of around 20 seconds before impatience and irritation set in – but software is improving and the connecting servers on the Internet are increasing capacity to meet demand.

No one is yet suggesting the perfect solution is in place, but many banks like Citicorp and Barclays as well as credit card transaction-processing businesses such as First Data have vested interest in building effective data and message protection systems that can provide as least as much reassurance in the virtual world as they do in the physical.

*Biometry*

Biometry is potentially the most secure level of message security. It involves using some unique aspect of the human body such as fingerprints, voice recognition or verification of retinal patterns. Special biometric software has been developed that would come up on-line during the ordering process. Customers, for example, would sign the screen in their normal signature using a digitising pen. The software would assess the speed of writing and stroke order among other keys to verify the signature.

Biometry is still in its relatively early days. There is no accepted global standard and a number of software companies like Peripheral Vision in the UK are developing their own unique approaches. As a result, adoption is still expensive and few commercial companies to date see this approach as their key. But again in the coming years software is predicted to improve, costs will come down and for sure biometry because of its ultimate simplicity and user-friendliness will find a place in the electronic commercial world.

*Firewalls*

Firewalls is the last of the main message-protection devices to discuss. The idea is that a core network should not be connected to the public internet without some form of protection. In other words it should be possible to put a 'wall' between the public Internet and private access. A firewall itself is a term to describe the computer or router that's specifically installed to filter messages and not allow them to pass from public to private access or back without verification or authorisation. Typically it allows insiders to have full access to the outside while controlling access from the outside in.

There are many computer companies now offering firewall services and it's quickly become a very sophisticated means of protecting message security and keeping hackers out. It's endorsed by the National Computer Security Association as currently the most effective way to protect data and message. The equipment can be quickly installed and the time delays in data transmission through the firewall filter are now negligible. Firewall systems are now commonly put in together with encryption providing particularly secure environments. Indeed a British company, Janus Software, claims to have now developed a totally hacker-proof security system. It recently put up a £10,000 reward to any hacker who could break in. The challenge attracted 80,000 hackers from all over the world. Its challenge ended in March 1998 and none had succeeded!

**Payment security**

In addition to the various initiatives in personal privacy and message protection there has been significant progress in providing payment security, especially with reference to credit cards.

IBM have pioneered a global standard called SET (Secure Electronic Transactions) which after a few early hitches is now becoming more widely adopted. In Europe a consortium of 38 banks are encouraging its use. They are led especially by Commerzbank with Kardstadt the German department store retailer, both of whom are far advanced in their trials around electronic commerce. The consortium is working closely with Visa and Mastercard and together they are promoting the service to all their members.

SET, like a rival approach called SSL (Secure Socket Layer) uses 128-key encryption to scramble the credit card data. It creates a secure channel to prevent third parties on the network from being able to tamper with payment information passed between the customer, the bank and the merchant. SET and SSL are expected to be fully 'debugged' and operational during 1998. Microsoft and Netscape seem to be moving to build in either SSL or SET into their browsers and many commercial web sites have expressed an interest in linking up.

The disadvantage of this approach is that 128-key encryption, while legal in the USA, is discouraged in a number of other countries including the UK, where government fears its own systems couldn't break into and therefore control that degree of sophisticated encryption. Nevertheless Barclays Bank, for example, in the UK is trialling SSL on its web site Barclay Square to provide its customers with a secure method of paying by credit card over the Internet. Since its launch Barclays claims 'it has had no fraud whatsoever'.

Payment security had been shown in the early days of Internet commerce to be a major concern for consumers and there is research to show that it has been an inhibiting factor discouraging shopping on the Internet. However, what is emerging is that technologies like SET and SSL, combined with initiatives in areas of data privacy and message protection, are creating safe and secure environments. If anything there will be more security shopping on the web than with traditional shopping methods. At the start of 1997 the *Financial Times* predicted, 'the internet will be safe for commerce within eighteen months'. *Business Week,* in its regular surveys of what is happening on the Internet, endorses this view but points out that as with many security systems, 'self-help' is also important and there are steps it also advises would-be electronic shoppers to take. These include some common sense items as 'keep security software current and download browser security updates when offered', 'don't lend your password to someone else to use', 'choose a bank like Wells Fargo that will cover losses from any cyber break-ins' or 'choose providers like AT&T that guarantees to cover any fraud for users of its services'.

If a consumer is still not comfortable then there are yet further alternative means of payment being developed which provide their own security and safety reassurances.

One of these is 'digital cash' or 'electronic cash'. It's intended to be a new form of tender that can be passed from buyer to seller almost like real cash. The electronic cash company is either linked to a bank or is part of one and it provides the consumer in effect with a digital current checking account converting the cash in the account into digital information and creating a kind of electronic purse for shoppers to use to buy goods and services. The consumer then purchases using an assigned digital account number which is suitably encrypted.

'Digi cash' or 'e-cash' providers claim it will especially interest consumers because all transactions can be anonymous: 'Without the spender's consent, no one should be able to trace who paid whom for what, the spender is the only one with the digital account number.' There were some 30 pilot schemes being tested through 1997 and 1998 from sophisticated environments in the USA through to places as far-flung as China and Zambia.

One particular advantage for digi cash or e-cash is that it lends itself more easily for 'micropayments', enabling consumers to pay for small items easily without the complexities of credit card data management and the processing overheads attached to it. For example, people may want to download one song from an album or access a new food recipe or a page of information on a tourist resort. The price might be cents not even dollars and electronic commerce requires a simple payment system to facilitate that.

Yet another approach is the smart card. This is a plastic card which can also act like a debit card. It can be loaded up with money. It can be inserted into the ATM and charged up with, say, $500 from the bank account. When consumers then want to place a shopping order electronically they can insert the smart card into a reader in the computer which together with a password or PIN authenticates the card and at the same time confirms to the 'cybershop' that the Bank will honour the amount to be paid.

SmartGATE is one software company now promoting the use of Smart card. They work with retailers to provide customers and employees with personalised PIN number smart cards which operate with software installed on the home or office PC. There are a series of verification and authentication steps built into the software, all the data that is transmitted is encrypted and there is a firewall that can be additionally installed to provide further protection.

Smart cards are widely used in France today where they function mostly as credit cards with some 20 million cards in service. In Germany, Siemens is testing out a 'super smart card' that in addition provides biometric fingerprint recognition. When the card is held with the thumb covering the lower right corner the card reads the actual fingerprint, compares it to the print stored digitally in the card and then clears whatever transaction is planned, whether withdrawing money, charging a purchase or even opening the door to your home!

No doubt some of these developments will stay in the lab but others for sure will become widely commercialised. There will be a number of payment methods available for Internet transactions, each offering different levels of reassurance and comfort to first-time or nervous shoppers. Certainly technologies are improving so rapidly that a secure commercial environment will be created and the *Financial Times* prediction at the time of writing looks readily achievable.

**General regulation and control**

At the moment there is uncertainty about legal jurisdiction in cyberspace. The legal status of electronic copyright, for example, is vague, as are the legal and practical issues surrounding on-line exchange and electronic cash. There is also no global authority to enforce rules on the Internet or possible sanctions for those who break the codes of practice. Yet these are issues which national governments, the United Nations, the OECD and trade associations are all acutely aware of and they are actively developing plans to manage electronic commerce more effectively either by regulation or voluntary code.

While suppliers of Internet services are very aware of these executional problems they have nevertheless spent the late 1990s focusing on getting themselves wired, establishing their web sites and exploring the market potential the Internet can bring. As the market grows and supplier experimentation matures so they will inevitably also turn their attention to the rules and laws of commerce and set about promoting more effective control and regulation.

In the meantime the US government has been setting the pace in investigating ways of developing the new rules and regulatory regime. For example in 1996 it established the Information Infrastructure Task Force which among other things has made recommendations on amending copyright laws to specifically cover electronic transmissions. In addition the WTO (World Trade Organisation) has set up working parties to make recommendations on commercial rules including reviewing the extension of individual governments' rights of jurisdiction from the physical into the virtual world. The French government has also tried to attack this problem by devising a central registration system for all who want to supply or use the Internet. As a more hard-hitting mechanism the FBI has put together its own task forces to track down hackers, on-line pornographers or suppliers who try to avoid taxation. Their goal is to ensure such individuals or companies get charged with criminal offences.

Another approach to improving regulation and control is the development of on-line communities that provide their own rules and securities. Companies with common interests could cluster themselves together. For example, auto parts suppliers might band together in an exclusive network serving the auto industry. Investment management firms may decide to sell their services only through distributors who have signed up to the same rules of commerce. Universities could choose to distribute course details only within their own selected network and effectively establish an extended intranet to groups of dedicated linked users. Banks, insurance companies, computer manufacturers could all set about doing the same thing.

Where common or mutual interest is insufficient, individual companies could turn to third parties who could choose to develop the on-line community and put in place a satisfactory framework for doing business. For example America On-Line (AOL) currently sells access to the Internet and provides a forum in which to advertise, buy and sell or supply information. In effect AOL has already established a well-regulated and well-maintained environment which does have a set of rules and creates a secure regime where basic laws of commerce are expected to be adhered to.

AOL and others like Microsoft Network may not see their role as law enforcers but they do in effect provide that value-added service. That need for control and regulation may well become a key growth driver for AOL and others as providers and shoppers are likely to become increasingly comfortable shopping in what is presented as a reputable and familiar environment. As this value-added role becomes more widely recognised we can expect others, especially the telecommunication companies, to step up their profile in managing access and offering secure 'community services' on the Net.

\* \* \*

Business has moved a long way since the Internet first went commercial. Since the early days when surveys showed significant numbers of executives weren't even sure if the Internet had any application to their organisation, companies have rapidly got more aware and sophisticated. Many can now see the market potential and want to exploit it. In chasing that goal they are themselves putting pressure on software and hardware operators to help make commerce on the Internet as smooth and as efficient as possible. Investment and trial in developing the infrastructure is moving at a fast pace and many more enlightened commentators now recognise that security will quickly cease to be an issue. 'Lack of security is no longer a barrier,' conclude US researchers Aberdeen Group, 'the debate is now quickly moving on to focus on improved bandwidths and delivery logistics. There is now sufficient security protection in place that especially where the provider is a recognised brand name there is now little perceived risk.'

Only the most reluctant electronic shoppers may still be concerned. But even now there are plenty of 'frenzied copers' and 'experimenters' out there wanting the convenience and the new forms of service. Their motivation is such that they will make the trade off between the new shopping advantage vs any lingering questions of payment security and will likely be quickly reassured there is sufficient protection and assurance built in.

# 12 How Can Manufacturers Respond?

In previous chapters we considered the retailer's dilemmas in facing up to the electronic shopping challenge. Now we turn the spotlight on manufacturers and suppliers. What will be the impact on their established ways of doing business, what are the threats and risks, is there an upside and what strategic options do they have moving forward?

In fact, they have a unique window of opportunity. They have a once-in-a-generation chance to truly connect with their end-user customer base. They can look to break their dependence on their retail customers. They can exploit the new technology and consumer interest by dealing direct – themselves. They can establish their own sales, marketing and home delivery contacts. They can disintermediate those who have for the last 25 years dominated them and the supply chain.

For the first time, manufacturers have a choice in how they distribute. They can decide that their future business *does* still lie with their existing retail customers. Or, they can decide to develop their *own* consumer supply lines. Electronic commerce creates this strategic crossroads (Figure 12.1). Down the first path may lie continued dependency but existing custom and business is preserved – at least in the short term. Down the alternative route lies adventure, new revenue potential, a platform for the future but also risk – risk that the retailers will see the manufacturers' ES activity as a direct competitive threat and cut off distribution. While electronic commerce is in its early phases, manufacturers can duck this issue. But as it develops it will come to confront them head-on and tough decisions will need to be made.

In addressing this challenge, all manufacturers will be acutely aware of the history of their relationships with their customer base. They will have seen their early dominance in the 1970s continuously undermined by their more sophisticated, larger, demanding distributors. They will be frustrated by relationships which are often confrontational and aggressive and despite sometimes warm words rarely move beyond periodic fights over price and discounts and rebates. They will wonder how they got themselves into this position where the retailer customer has

**Figure 12.1** Manufacturers at a Crossroads

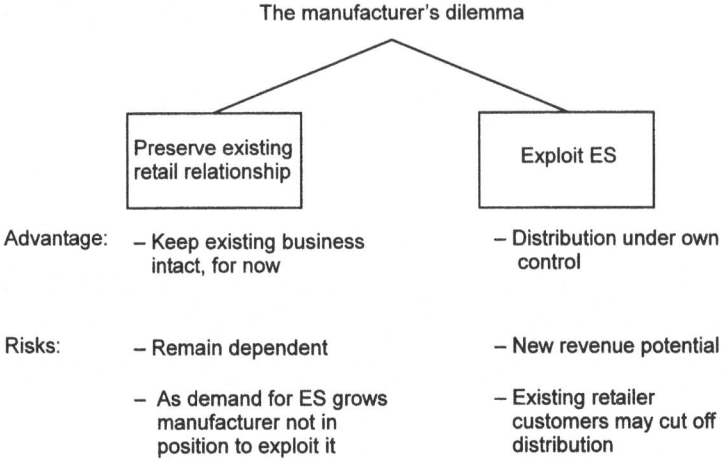

so clearly wrestled the balance of power from them. They will want to challenge whether they do have a profitable future continuing with these relationships. They will want to explore just how much flexibility they do have in pursuing alternative options. Do they have the competencies and the strength of purpose to develop their own direct operations? Do they have or can they acquire sufficient market clout that they can be sure consumers will continue to clamour for their products – wherever they are distributed or located? Do they seize this opportunity now, do they wait and see or do they conclude that the risk of losing existing business is too great?

To understand how manufacturers can best respond to these issues we must first review in more detail the historic baggage they are bringing into this debate. In the early days, in the 1970s (Figure 12.2), when manufacturers' brands were generally dominant, they could virtually dictate terms to their distributors. In the supermarket sector, for example, which has been a hothouse environment for the supply relationship, companies like Heinz, Campbells, Procter & Gamble all had well known household brands that had strong franchises. Their technologies were still relatively immature so they could establish a stream of innovative new products which continued to keep their proposition to the market fresh and stimulating. In some countries and sectors there was even a system of 'retail price maintenance' where the manufacturers could print on the pack or otherwise control the price the product was sold at, thereby

**Figure 12.2** Retailer–Manufacturer Relationships

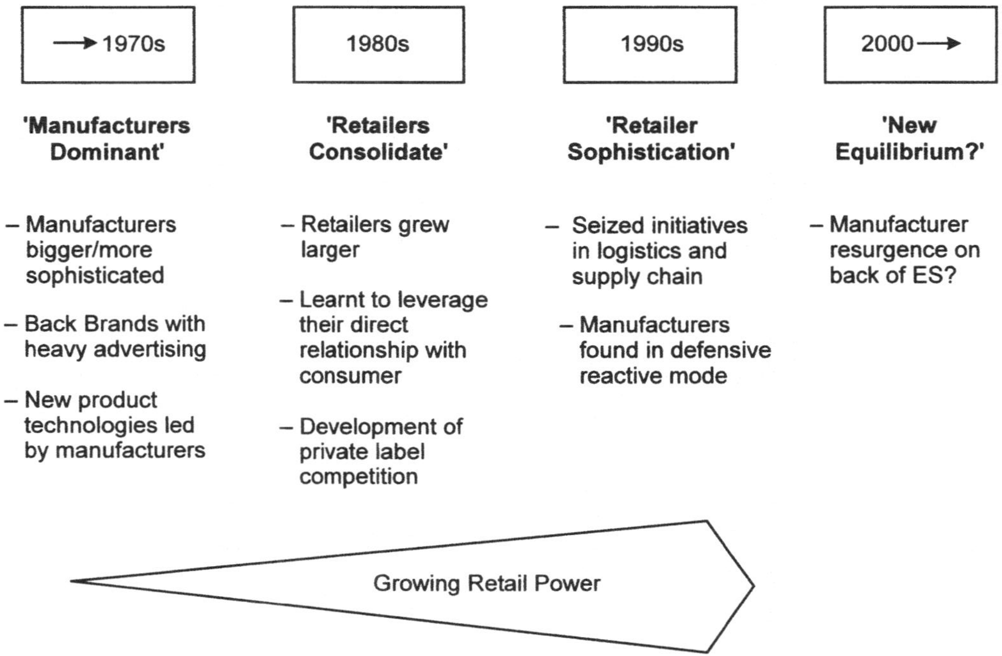

| →1970s | 1980s | 1990s | 2000→ |
|---|---|---|---|
| **'Manufacturers Dominant'** | **'Retailers Consolidate'** | **'Retailer Sophistication'** | **'New Equilibrium?'** |
| – Manufacturers bigger/more sophisticated | – Retailers grew larger | – Seized initiatives in logistics and supply chain | – Manufacturer resurgence on back of ES? |
| – Back Brands with heavy advertising | – Learnt to leverage their direct relationship with consumer | – Manufacturers found in defensive reactive mode | |
| – New product technologies led by manufacturers | – Development of private label competition | | |

Growing Retail Power

175

176                                    *e-shock*

guaranteeing their profit margins and locking them in at attractive levels. In contrast, their then distributors were relatively small, fragmented and lacking the know-how to exploit their contact with the end-consumer.

Through the 1980s all that began to change. Demand in most developed economies was buoyant. Retailers began to expand rapidly building new stores and adding space with a frenzy. Part of that expansion also brought with it the inevitable wave of mergers and consolidation as some tried to achieve winning positions more quickly. As these developments bore fruit so retailers also began to stand up to their manufacturer suppliers. Suddenly they realised that they often accounted for c.10% of a manufacturer's business and they found their threat to clients were usually accompanied by a gradual giving-in to whatever demands the retailer was trying to impose. Power changed hands. Within a few short years a handful of 'super retailers' began to emerge. They backed up their new-found clout by investing in systems like EPOS that gave them greater knowledge about their customers' buying habits, which product lines were selling especially well and which were below target, and linked to DPP they could tell which were adding profit on a net profit per sq. foot basis and which weren't. New-found IT power, new market awareness, increasing breadth of range and choice in

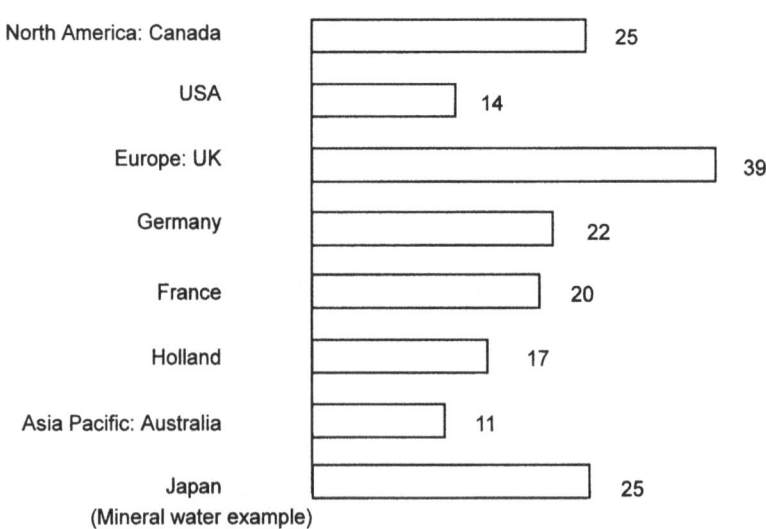

**Figure 12.3** Private-Label Grocery Penetration (%)

*Source*:  Wileman and Jary, *Retailer Power Plays*.

store, smarter environments all combined to put them in the driving seat topped now by their own brands. Why be dependent on Coca-Cola or General Mills or Kraft when they could provide similar quality at lower prices to what had now become 'their' consumers? By the mid-1990s retailers' brands had taken over from manufacturers in many sectors in many countries (Figure 12.3).

As we now approach the end of the 1990s, retailers are as strong as ever. They are the power base. Manufacturers remain on the back foot. Observers must feel they're at a boxing match watching two heavyweights slugging it out. As the contest has gone on it's become a wretched spectacle. The result looks a forgone conclusion. The weaker fighter is punch-drunk, lurching around the ring looking dazed, struggling to stay on his feet, hoping for some miracle intervention. The stronger fighter is dancing, cock-a-hoop, throwing punches at random – some looking suspiciously below the belt. Let's look at some of the photos from the fight so far:

- 'Sainsbury has humiliated Coca-Cola, the biggest and most heavily hyped brand ever, by grabbing 60 per cent of in-store cola sales with its own private label brand. Sainsbury is now launching its own brand of luxury ice cream to rival Häagen-Dazs and Ben & Jerry's. It was not more than a year ago that Sainsbury's own washing powder brand Novon stole market leadership out-manoeuvring mighty Procter & Gamble and Unilever.'

    (*Times report, 1996*)

- 'The war is still on between retail-owned products and manufacturers' products in Japan. Japanese manufacturers have been facing the onslaught of private retail brands prompted by sluggish domestic demand. The easing of retail laws has increased the number of large scale stores and retailers with extensive networks have taken advantage of their distributor power.'

    (*FT article, 1996*)

- 'Unilever used yesterday's AGM to vent its irritation at the emergence of so-called retailer copy cat brands which closely imitate manufacturer brand names and packaging but undercut on price. Unilever has joined with other industrial groups such as Coca-Cola in an attempt to persuade the UK government to prohibit what they regard as unfair competition.'

    (*FT, February 1995*)

- 'The trade philosophy in France is a hard discount mentality. It is antagonistic to the manufacturer. Margins are being pared to the bone and there is no sign of a let-up.'

(*McKinsey report*)

Not surprisingly, against this background, manufacturers generally are extraordinarily nervous at doing anything that might make their adversary fight even harder. They're very aware that developing their own direct links might be taken as a red rag to a bull. Yet they see the enormous opportunity ES can present. Is there a way out of this dilemma? Can they find the inner resourcefulness, as the bell rings for round 12 in the boxing match, to get back into the contest, work out a way that lets them pursue their own tactics and at least get themselves back to all square. They don't need to find the knock-out blow – win:win would do.

More prosaically, can they manage the conflict of distributing through retailers and direct to consumers at the same time? Aren't there any success models out there to draw on? Some manufacturers must have managed this already, if not yet in an electronic context then nevertheless dealing with similar market pressures? And indeed, thankfully for our hapless pugilist, there are some pointers and lessons learned. Companies like Procter & Gamble, Sony, Nike, Mars, Gillette, Seagram, IBM, Kellogg's, Mattel, Unilever and others – companies operating in many different retail sectors – have demonstrated that it is possible to manage retail customers successfully. They have shown there are ways to pursue their own goals and initiatives while still building and sustaining a successful business with retail distributors.

Coca-Cola provides one immediate illustration of a manufacturing company that sells simultaneously through different channels and has learnt to live successfully with any channel conflict that might result. As the company has grown, Coca-Cola has developed business through grocery multiples, discount stores, small neighbourhood convenience stores, bars, restaurants, delis, theatres, take-home fast-food outlets, vending machines as well as even having some showcase stores of its own. Yet Coca-Cola has persisted with its competing channel customers and over time has persuaded them to largely accept this. It's acknowledged as an inevitable condition of the market place.

What have Coca-Cola and other successes done? How have they managed the inevitable friction when they do open up a new channel of distribution? How they have dealt with the possibility that one channel's

## How Can Manufacturers Respond?

sales will cannibalise the others? How have they responded to existing distributor's initial threats to delist the product and stop distribution?

Researching and interviewing with these companies it becomes clear that their key lies in segmentation. Their route forward is to clearly differentiate between the channels and establish different value propositions. The learning is that separate channels exist because by definition they are responding to customers who have different needs. One wants to buy a multi-pack of Coca-Cola as part of the weekly supermarket shop, the other wants the convenience of a single can purchased on impulse at a vending machine on a street corner. The answer is to understand those different customer needs – profoundly, develop the differentiated propositions and market them in that way. Do this persistently, accepting some initial kickbacks until the market place accepts it as a necessary and integral part of the market scene.

There are in fact three segmentation options, each is illustrated with brief case examples:

1. Product Segmentation
2. Territorial Segmentation
3. Complementary Segmentation

### 1. Product Segmentation

- Black & Decker has developed three distinct product ranges each targeted and tailored to different channels and set up to provide a differentiated proposition. Black & Decker uses its Quantum brand at the heavy DIY end, the Black & Decker brand for the major multiples and De Walt aimed at trade buyers and contractors. While this does create significant product-line complexity for Black & Decker it has helped manage potential conflicts. For example, different promotions and incentives are made available but because they're on different brands the scope for comparison or conflict between channels is much reduced.
- In a similar way, Levi's established Britannia jeans and the Levi's brand in major multiple outlets, Dockers through department stores and speciality retailers and Slates for upmarket stores. To strengthen the product segmentation Britannia was targeted at the lower price end of the market, Dockers is mid-range but its image is young and fashionable, while Slates is targeted more to the 30+ age group.

## 2. Territorial Segmentation

- Nationwide Insurance in the USA has long relied on its network of agents to build its business. However it has seen the potential of direct consumer links. It has evaluated how to build its own direct channel while preserving the integrity of its agency network and their business potential. Nationwide's solution has been to split the USA geographically. In a number of states the agency network was especially strong and rather than risk cannibalisation, Nationwide has kept those as 'agency states' and is working to reinforce the business there. However in other states, Nationwide either had no or limited agency coverage. It is in these states where it is building its direct network, learning what can be achieved, whether the two approaches to market can work in harmony and what is their long-term potential.
- Kodak, when it was in the photocopier business, had a long history of selling through distributors. But relationships were often unsatisfactory. In many regions there were several rival distributors covering the same territory. Each demanded more favourable business deals and invariably assumed, even without evidence, that Kodak must be treating one more favourably than the other. There was continued conflict and sales were being jeopardised. Kodak determined to resolve this. Its solution was to select the best distributor for target groups of customers. It picked one target distributor per region. While there was some initial sales fall-off from disappointed distributors, Kodak gritted its teeth and by persistent application gradually built its chosen distributors into dominant but differentiated sales channels.
- Now ASDA, the UK supermarket chain, has announced plans to set up its own electronic ordering and delivery service *but* it plans to establish this in territories where it has few if any stores. In doing this it minimises cannibalisation, reaches new customers it could never normally reach and attacks the home bases of its competitors!

## 3. Complementary Segmentation

- Coca-Cola did have significant problems from its established retailer outlets when it started installing vending machines on a wide basis. But Coca-Cola had also conducted detailed consumer research that showed that vending machines sales could actually *add* to the total Coca-Cola brand and market. Sales could be incremental rather that cannibalising existing outlets. The vending machines did serve the same consumers, but did so on different occasions. Tests had shown

the vending operation could actually benefit all parties. It was a classic win:win.
- Similarly, Nike ran into problems when first opening its own Nike Town stores. But it too had carried out detailed consumer surveys that clearly demonstrated the role the stores could play. They were perceived as reinforcing the brand name, as showcases that could also generate increased demand and sales in the established channels. Again the case was put forward that this could complement not cannibalise. Subsequent market success proved the point.
- The major pharmaceutical companies can also provide a lesson. They have been switching sales and marketing effort from traditional independent pharmacies toward the large HMOs and mail-order drug operations. But they have taken the view that both channels need to and can coexist as complementary sales operations. With that goal they have developed distinct but separate marketing support packages which do recognise where the future volume sales will lie but do also provide specific support at a local level specific to what the independent pharmacies need. While initial coexistence was not easy, the independents – who stood to lose the most – have come to appreciate the pharmaceutical companies remain committed to them and are backing that up with sales and marketing action.
- Other examples come from Sears, who continued to provide discounts, rebates and business support to its distributors in small towns even though its business had begun to shift toward the major malls and outlets. GE, too, had originally depended for its business on a vast number of small independent dealers and even as its business shifted away to larger outlets it continued to provide sufficient support in financial deals but also through its salesforce. The same dedicated and complementary approach to different channels has been taken by pet food suppliers as their business develops away from independents to the larger outlets like PetsMart and also by Avon who continue to base their business on their large number of sales representatives while targeting new and different types of consumers via their Internet operation.

Channel conflict can be managed. It's not easy. It takes time and persistence. It requires a clear value proposition that reflects a deep-rooted understanding of customers and their differing needs. It involves a focused segmentation and a differentiated sales and marketing approach. The channels may be naturally complementary so that all will gain or they may be more directly in conflict. Whatever the situation, the onus rests on the manufacturer supplier to find the win:win where all parties can ultimately benefit. It's an education and acclimatisation process in which manufacturers and their distributors can get used to a new set of market variables, readjust expectations and learn to operate effectively with the different challenges.

Which manufacturers are best placed to succeed in a channel conflict situation? What steps can be taken to ensure the segmentations that get identified can be effectively marketed? How can they ensure that existing customers do gradually adjust and accept the new market conditions? The research has shown that there are some ten key criteria a manufacturer must aim to score well against. Each criterion represents an organisation and market state that manufacturers must aspire to if they dare to succeed in the channel conflict game. The stronger the position the company can put itself in the market the more able it will be to arrive at a satisfactory end game.

The ten key criteria are set out as questions in Table 12.1. Current market strength and flexibility to manage the ES challenge can be measured by scoring each question 0–10. A strong Yes answer scores 10, a strong No 0. A high positive score > 70 would indicate high chances of success. Lower scores would raise a cautionary note and suggest that *before* embarking on a potential channel conflict, the company has work to do to get its products, customer understanding and market position up to strong enough levels that it can take on the challenge. That break point < > 70 is critical. If a company does set up its own direct consumer links and takes on a channel challenge without sufficient market clout it will likely fail. If it recognises its weaknesses beforehand then it still has time to build a sufficient platform so it can succeed. In this regard, electronic commerce may be the trigger that gets the manufacturer to finally deal with some of its sales and marketing issues that have bedevilled it but have been allowed to continue. It may have too many products or brands to support, it may need alliances in order to access more marketing funds, or improvements in its product technology, it may never have adequately segmented its customers and neglected to respond sufficiently to their underlying needs. Most companies will be able to intuitively score the ten questions set out and know immediately for themselves what are the key issues they need to address.

Leading manufacturers have driven hard in recent years to reach a position where they could score highly, where they have got the market clout. For example, Philip Morris, Campbells and Unilever have been aggressively rationalising their product portfolios to focus down on their major brands and give those the attention and investment that can get

**Table 12.1** 'The Manufacturer's Test' – has the manufacturer the strength and flexibility to manage channel conflicts?

| | |
|---|---|
| 1. Does the company currently sell successfully through different channels with significant sales in each? | _____ |
| 2. Is there a rigorous process institutionalised in the company for researching customers and precisely understanding their different needs? | _____ |
| 3. Is that knowledge effectively exploited so that there are truly differentiated value propositions for each channel? | _____ |
| 4. Is that differentiation also translated into different and targeted sales and marketing programmes responding to segment needs? | _____ |
| 5. Has the company been able to establish strong collaborative partnerships with its customers vs a relationship of confrontation and mistrust? | _____ |
| 6. Does the company have a strong brand portfolio where most products are No. 1 or No. 2 in their sector? | _____ |
| 7. Would industry observers rate the company's sales, marketing and channel management skills as core competences? | _____ |
| 8. Is the company building off a relatively solid platform of rising sales and stable margins? | _____ |
| 9. Does the company have a history of ring-fencing sufficient sales and marketing funds and investing consistently and persistently? | _____ |
| 10. Does the company have the leadership at senior management to take bold market place decisions and see them through? | _____ |

them to No. 1 or No. 2 market positions, and keep them there. If consumer demand is strong enough then of course they have much greater flexibility to explore new distribution routes to the end-user. As another example, computer hardware manufacturers such as Hewlett Packard, Compaq and IBM have put substantial efforts into improving their supply-chain capabilities. They have put themselves in a position where retail distributors *want* to deal with them, not just because of their brand but because they can more or less guarantee stock availability, consistent supply and product reliability. Other companies like P&G, Johnson & Johnson or Levi's have focused on the pull-through effect of their brands making them a 'must stock'.

There is no doubt, then, that a number of manufacturers in different sectors arrive at the electronic shopping starting post with all the right qualifications, competencies and experiences. Of course e-commerce itself is new and different, it will pose its own unique challenges but the underlying market and channel issues are at the end of the day the same. It will all come down to being able to pass the Manufacturer's Test with a high enough score, establishing the 'channel conflict-management confidence' and then rigorously pursuing an effective segmentation strategy that is set up to meet different consumer needs with demonstrably different value propositions.

To date, only a few manufacturers have risen their heads above the parapet and made any public pronouncements about their ES plans and their preparedness to manage any channel conflicts that arise. But as the electronic channel to market grows and matures, as is predicted, then expect more of the leaders to come forward with strong market propositions ready to see them through. Unilever has already declared its interest and this excerpt from a recent interview with its Chairman provides a suitable resume of both the manufacturer's dilemma but also its need to come to grips with it and move its business on:

> 'Unilever is one of the few multinationals to go public and be proactive about its on-line intentions. It has announced a radical diversification into new service-related methods of channelling its foods and personal products to the consumer . . . areas being examined include home selling on the personal care side where products include Elizabeth Arden cosmetics.
>
> Niall FitzGerald, Unilever's head, said such arrangements needed to be studied: "in both food and personal products we are seeing the emergence of alternative channels to the consumer . . . unless we are careful we will find ourselves boxed in to what is sold through a consolidating block of retailers. So we need to consider ways to broaden the channels through which we can reach the consumer."'
>
> *(Times, 1997)*

# 13 Manufacturers' Ten Strategic Options

As manufacturers reflect on the new electronic shopping challenges, must they inevitably transition their business to embrace it? Can they sensibly resist it, eschew building up their own direct operation and stick with their existing distribution arrangements? Is there a half-way house that keeps a foot in both camps and gives the flexibility to move one way or the other as the market develops? What are the basic strategies that a manufacturer can choose, what are the options?

The testing and research have identified ten strategic options (Table 13.1) that a manufacturer could pick from. They fall into different categories. The first two are 'partnership-based' and assume future success can be derived from getting even closer to existing physical retail customers. The second two are 'pull-based' and care less about channels of distribution, simply focusing on making the products so attractive that they will be demanded and sell through any and every route to market. The next two take the first steps down the ES path. The final four strategic options involve a more significant electronic commitment all the way through to a 'fully direct' transition.

Table 13.1  Manufacturers' Ten Strategic Options

| | | |
|---|---|---|
| **'Partnership-based'** | 1. | Woo the retailer |
| | 2. | Move to private label |
| **'Pull'** | 3. | Technology-led |
| | 4. | Brand-driven |
| **Electronic Information-based** | 5. | Information only |
| | 6. | Forming a club |
| **'Own Direct'** | 7. | Treat as another channel |
| | 8. | Set up as separate business |
| | 9. | Band together |
| | 10. | Go fully direct |

Determining which is most attractive is going to be something for each manufacturer to evaluate and decide on an individual basis. The decision making can be guided using the two tests that have been described in Chapters 5 and 12:

- The 'ES Test' will provide a clear indication of the product's attributes, its susceptibility to the forces of electronic shopping and the degree to which target consumer groups are likely to be responsive to ES initiatives and demanding of that means of selling and distribution.
- The 'Manufacturer's Test' will help determine what market strength and flexibility the company has. The greater the market clout the greater the choice and the more likelihood routes that break dependence on existing partnerships could be made to work.

**Figure 13.1** Picking from the Ten Options

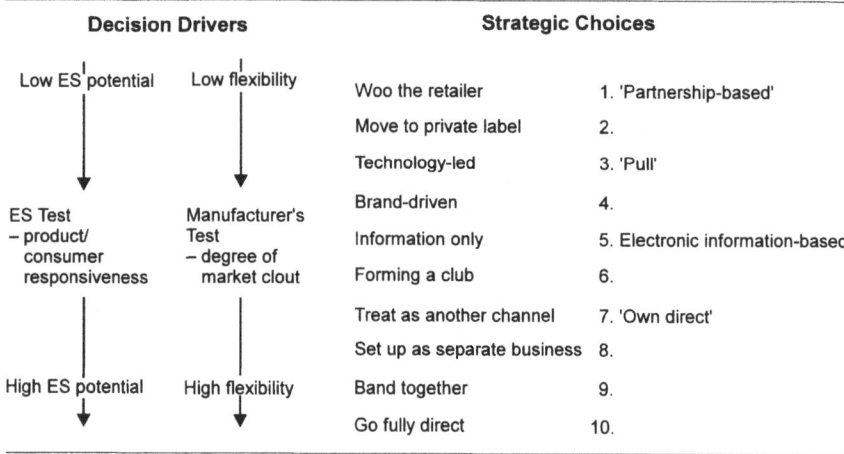

Scores on these two tests will influence choice of strategy. They will help identify which of the ten options give individual manufacturers the best chance of success (Figure 13.1). If there's high ES potential, then only a high 'Manufacturer's Test' score, for example, will make the full range of options available for the manufacturer/supplier in the market place.

Before going further in comparing the attractiveness of the different options, each can be illustrated and described in more detail to provide a clearer definition of what the different choices entail.

## 1. 'Woo the Retailer'

The benchmark here is the Procter & Gamble Wal-Mart relationship. These two organisations don't just talk about partnership. They've gone beyond co-operative marketing or 'joint action teams'. In many respects they have locked together into one business system (Figure 13.2).

**Figure 13.2** Sharing Responsibility

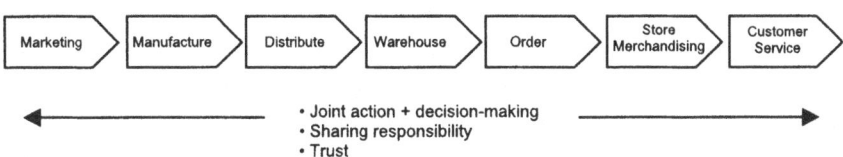

P&G and Wal-Mart now share responsibility for the supply chain, second staff to each other's operation (Wal-Mart has some 50 P&G people permanently located at its facilities) and decide which is best placed of the two to take the lead and accountability for different supply initiatives. For example, P&G is 'trusted' to do the store re-ordering of its products and a process has been set up to agreed parameters which instantly replenishes shelves and warehouse stock levels, triggers the invoice and causes funds to be transferred – all automatically. Costs are reduced, errors minimised, stock-outs unheard of, the relationship is vigorous and each is responsive to the other's needs. They already deal electronically and it is easy to imagine the two partners now jointly investigating and exploring how to best reach the new electronic consumer in a way that is to their mutual advantage.

The benefits of collaboration can be considerable for both parties. Recent research by Professor Kumar showed that where retailers and manufacturers operated with a high level of trust sales generated were anything from 11% to as much as 78% higher, compared to situations where there was confrontation and antagonism. As Kumar points out:

> 'Trust creates a reservoir of goodwill that helps both parties fulfil their potential ... sharing information, investing in each other's business, customising systems, dedicating people and resources ... and capturing economies of scale and efficiency in working together.'

In P&G's case its 'wooing' of Wal-Mart has lead to it generating more than $3bn in annual sales together, which is about 10% of its total revenues. The profitability of those sales has been 'significantly enhanced'.

The automotive industry offers another illustration of how partnerships generally can operate to mutual benefit. Unlike the P&G Wal-Mart situation, examples from the automotive sector typically deal with smaller manufacturers and show that even if there is an imbalance in size between the parties nevertheless a meaningful partnership can be forged that works to each other's advantage.

Pressac is good example of a small electronics manufacturer based in the UK with a turnover of less than $100mn yet successfully supplying leading automotive manufacturers like Ford and GM around the world. Pressac specialises in electrical components such as lampholders and wing mirror heater elements and has dedicated itself to achieving excellence not just in product quality but in its commitment to making its key customer relationships work. Pressac has the advantage of supplying to a sector which is very supplier-oriented and which has long recognised for itself the value of close relationships. For example, Nissan's stated strategy is

> 'to achieve our objective through long term close working relationships with a limited number of suppliers and to develop that supplier base so that together we are capable of achieving Nissan's development plans world-wide.'

Retailers of fast-moving consumer goods are generally some way behind the automotive companies in cementing this level of partnership and realising its benefits, but there are a few pointing the way. One such company is Marks & Spencer, who not only have sophisticated manufacturer partnering arrangements but show that success can be achieved by small manufacturers as well as large. For those manufacturers who commit to them and demonstrate their product excellence, M&S can be a supportive and caring distributor partner, keen to ensure its manufacturers make a decent enough margin and are not so squeezed that they cannot sustain their own business and continue to build it. Successful suppliers to M&S find they are invited to join a club which involves among other things being given a key card to enter M&S's head office encouraging them to drop in to discuss any small issues. There is also joint product development and joint investigation of original source materials, participation in internal store operation meetings and guaranteed access to the 'boss' if things aren't working well.

Achieving this mutually beneficial state is not easy, of course, and requires years of proactive investment and effort especially now from the manufacturer to overcome initial scepticism and show that previous adversarial dealings are decidedly a thing of the past. Manufacturers who choose the partnering approach and select this strategic option to deal with the ES challenge will have to be persistent and dedicated if they are to reap the benefits. They'll need to work hard to get to the level of trust and collaboration so that future market developments are faced together and not with self-interest at the other's expense.

JE Ekornes, a Norwegian home furniture manufacturer, represents a good illustration of the effort and change required to get the benefits of this approach. In the mid-1990s it decided to review its European sales operation and was disturbed by a singular lack of progress especially in its French retail market. It sold in France through 450 distributors and quickly realised that its relationships with them were very poor and this was proving the major roadblock. Neither side shared information or trusted in the other's commitments and the brand and sales were suffering.

Ekornes decided to focus on a core group of distributors and turn that relationship with them into a true partnership. This meant changing the economics so that these distributors could make a satisfactory margin. To achieve this, Ekornes set about restructuring its position in the market. It awarded chosen distributors exclusive territories to remove the previous conflicts. It set up long-term contracts and guarantees of on-time supply. It established a jointly led process for improving the supply chain between the two. It involved the distributors much more heavily in its own business, getting them over to the Norwegian head office and factory, for example, to spend time there and understand their manufacturing operations. It also put in place joint system links so each could track order and sales progress. As a result of these changes, roles and expectations changed. For example, Ekornes redefined the role of its salesforce. Instead of just selling it gave them a stronger responsibility for relationship management and providing a more general level of business support. These changes and restructurings paid off. Sales tripled over three years.

There are other examples that can be drawn from Compaq, Goodyear, Sherwin Williams, Sara Lee and Philip Morris – all of whom have in the last few years made major strides, which have been well publicised, in achieving the benefits of partnership. Successful 'wooing' means breaking down the old relationship barriers, changing ways of operating and interacting and persistently signalling win:win intentions. Once major distributors and customers recognise that there is a long-term commitment to behave in this way, then enormous advantages can come from sharing the future together.

For the manufacturer who remains uncertain, even after the ES and other tests and checks, partnership is the most natural evolutionary step. It is most likely to maintain the status quo and is no doubt exactly the sort of strategic response any retailer wants to hear. The only risk, of course, is that if the retailer itself is either slow to respond to the specific ES challenges or otherwise finds itself disintermediated by more aggressive competition then the manufacturer, too, can find itself left high and dry and outmanoeuvred.

## 2. 'Move to Private Label'

The partnership route assumes that manufacturers retain the integrity of their product and brand proposition and even while collaborating on parts of the supply chain they nevertheless retain their independence and ability to determine their own future. However, when a manufacturer moves down the private-label path much of that independence and flexibility is surrendered. The manufacturer is now firmly locking its future in with its retail customers.

The principal benefit is the greater sharing in their retailer customers' success. Provided the retailers do well the manufacturer's sales and profits will prosper. But if the retailers do poorly the manufacturer, of course, has limited options and possibly few alternative sources of distribution.

What is the appeal, then, of this strategy? Why choose it if it involves limited long-term flexibility and has a high dependency level of risk attached to it? Those manufacturers most interested will likely be today's small–medium-sized companies who find their current brand product portfolio relatively weak vs the competition. Such companies will also typically find themselves in market sectors where the technology is mature, consumers see products as largely 'me too' and commodity-like, the sector is dominated by one or two large long-established competitor brands and retailers themselves have likely already introduced their own store label.

In such circumstances, smaller manufacturers are sub-scale and struggling to find the investments and other resources to stay in touch or attempt some meaningful differentiation. In these situations the arrival of yet another major new market place challenge such as electronic commerce may be one challenge too many to fight on their own. Just in the past ten years they will have already experienced the turmoil of globilisation, the arrival of new low-cost, low-price competitors from overseas, new technology raising the hurdle on supply-chain activities and efficiencies, pressure on margins as the lid is kept on inflation and consumers search for everyday low pricing and value. Now electronic commerce would appear to demand the opening of a new channel, more investment, new skills and possible head-on conflict with the retailer customers that they have for so long been struggling to keep on side.

For companies in this position moving to private label supply may now be the most attractive long-term solution that maintains shareholder value and has the prospect of increasing it. Work with, rather than in competition to, the retail customer, become integral and integrated into the customers' supply chain, tie future development into the customers' strategies – and hope that the customer understands its market challenges and has itself a clear view on how to deal with ES.

There are successful private label suppliers in every industry sector. Dewhirst, for example, is a clothing manufacturer which has long had established relations with Marks & Spencer and has prospered by building that relationship into a mutual dependence that would make it difficult (though far from impossible) for M&S to find an alternative source of reliable quality supply. Dalgety had for many years an established contract with McDonald's supplying buns and other food ingredients which had grown into an open understanding on ways of operating, profit margins and future product development. United Biscuits, the food and biscuits group, has been one of a number of companies that have developed a substantial private-label business but still managed to maintain its own product brand integrity.

Moving down the 'private label' path can certainly provide a meaningful route for dealing with the ES challenges. It forces and forges a sharing and collaborative response. But as with the 'partnership' route, it leaves the manufacturer much less room for manoeuvre if the retailer itself is outwitted and disintermediated.

## 3. 'Technology-Led'

As we go down the list of strategic options we start moving away from retailer-based initiatives. Instead, there are alternatives which are 'pull-based'. They demand of the manufacturer the capability to stand alone in the market place and build the strongest possible consumer proposition.

Being technology-led simply assumes that it will be the uniqueness of the product's performance, its advantage in quality, functionality and reliability, that will drive success. This approach carries with it an implicit disregard for channel conflicts. It assumes that consumers will so obviously and clearly recognise the product's differentiated benefits that they will demand the product – wherever it might be available. There will be a 'pull' for the product such that the manufacturer need not 'push' it through intermediaries. It will sell itself.

Such a strategy requires the manufacturer to build its core competencies around its R&D and ensure there is a continuous pipeline of new products and new developments that can keep a product's uniqueness and advantages firmly in place. This is especially demanding as it requires substantial investment and a deep-rooted commitment to keep that long-term spending in place even when short-term earnings are under pressure.

It has become an even more challenging option in the late 1990s. This is because product performance life-cycles are generally shrinking as imitators become more successful in quickly eroding any advantages. But also in a number of industry sectors technology has reached a level of maturity where there is limited further product improvement to be had. In cellular phones, the Walkman, watches, calculators, high-fidelity equipment, household consumer products like soaps and cleaners, basic financial services such as mortgages, pensions and insurance and in many other sectors there appears to be little scope for new technical breakthroughs.

Therefore any manufacturer considering this strategy must look hard at its technical skills and their long-term potential and applicability in the market place. For sure, a number of companies show it can be done. For example, Merck, with its commitment 'to be the pre-eminent drug maker through the best R&D capability', Toyota with its emphasis on superior product quality and reliability, Sony with its long-standing research and development in consumer entertainment equipment, Procter & Gamble and Unilever with their extraordinary efforts to keep innovating in an otherwise tired commodity soap and cleaner sector – all these companies and others have successfully built their businesses around technically superior products.

But as these same organisations now face the challenge of electronic commerce they must look over the next ten years or more and both question and reassure themselves that this distinctly 'pull' approach can continue to work even as consumer choice and sophistication grows and as channels of distribution fragment further. Those who remain determined to be 'technology-led' may nevertheless also want to combine this approach with one of the later strategies to be discussed and also set up some form of direct electronic communication to customers in support.

## 4. 'Brand-Driven'

Like being technology-led, this approach takes a 'pull' rather than 'push' route. It demands the creation of a set of emotional values, awareness and recognition that are so strong that the consumer will demand and search out the product, in whichever channel it is available. The approach requires all the emphasis to be placed on to the product and its branding and again, as with being technology-led, the commitment and funding must be unwaveringly in place to sustain success.

A number of companies have already established their brand-driven credentials. They come from many different industry sectors. So we see, for example, Coca-Cola and Pepsi soft drinks, Canon Copiers, Intel inside, DuPont Lycra material and Teflon coating, Citibank banking services, in fashion clothing the likes of Donna Karan, Giorgio Armani and Liz Claiborne and in sports equipment (the latest 'battle of the brands") Nike, Reebok, Addidas, Puma and others. For all these companies, and many others like them, their goal is to continue to build this pull brand so that even as new ways of shopping emerge, the awareness and desire remain strong. Such an approach can remain indifferent to the distribution challenges of ES, assuming that consumers and intermediaries will work out ways on their own to find, distribute and purchase the product.

Of course, this 'go-it-alone' approach has the risks associated with any such strategy and no company can afford to do this without at the same time being also close observers and watchers of their market place and of consumer desires and behaviour. If the competitive environment begins to shift such companies must be well placed to anticipate these changes and move in tune with if not slightly ahead of consumers' evolving needs. Only in that way can the brand stay lively and the pull approach truly work. (More detailed discussions of brand-driven strategies can be found in my book *Strategy in Crisis* (1997), which provides many case examples on how to effectively establish a branding or 'emotion-based' approach.)

## 5. 'Information Only'

This is the first of the strategies that begins to deliberately and explicitly explore electronic commerce. It simply takes the route of providing information only to would-be purchasers about the product, its features and then directs where it can be obtained. It's the starting point for most companies looking to develop and learn more about electronic commerce. Most organisations have already moved down this non-contentious path and set up their own web site. The initial applications have invariably been information-based without adding on transaction features. Digital Equipment Corporation, now part of Compaq, provides a good case study.

Eighteen months before actually setting up its web pages, Digital had begun exploring alternative ways to reach the market place and provide customers and prospects with information about its products and services. Its goal was to build some foundations to anticipate the coming convergence of the computer, telecommunications and consumer electronics industries. It had already developed a CD Rom demonstrating products and software and it was excited at the opportunities of interactive communication on-line and establishing its own direct consumer links without having to go through traditional distribution channels.

Digital evaluated its options against three particular criteria: (1) customer demographics – what sort of customers did Digital typically target, and how responsive would they be to on-line access and communication? (2) product characteristics – did their products lend themselves to description and demonstration electronically? (3) cost – as this was a new marketing approach, Digital wanted to manage its initial investment costs so that it would provide a learning environment, without putting pressure on getting quick payback.

The Internet appeared to Digital to meet all its early needs – a global reach, a suitable environment for its products and relatively low start-up costs. The sales and marketing department could immediately see opportunities to set out their products and show their capabilities, provide technical specifications but make it available only to those who wanted that level of detail, provide information, locations, telephone numbers of all its distributors organised geographically, list details of upcoming conferences and shows and invite customers and prospects to send in information and comment on products of interest and give their views on products' strengths and weaknesses.

As a result, as Digital developed its web site so it offered e-mail to encourage this two-way flow of information and reinforced that with newsletters and also setting up user groups to continue the exchange of data and experiences. From all this Digital could learn a lot about its products, its customers and its competitive strengths and weaknesses. Digital first launched its web site in October 1994. It contained among other things:

## Manufacturers' Ten Strategic Options

- New product announcements
- e-mail and user group facilities
- The ability to learn first hand about new software and hardware by logging on and actually 'test-driving' new systems
- Product brochures and specifications
- Interactive catalogues detailing all the products and where they could be obtained
- Links to related information to clarify new concepts
- Audio and video multimedia displays to illustrate particular product features and examples of usage
- Interactive features engaging the customer in a dialogue about product requirements or usage ideas
- Customer feedback pages.

Digital's web site was accessed 4 million times by over 100,000 customers and prospects in its first two years. We shall discuss in chapter 14 the marketing of web sites and how to go about attracting customers and prospects to it. But for the moment this Digital story shows how a company can get an information only web site to work very effectively as a business-building tool in its own right.

As it happens, Digital has gone on to build in transaction facilities to make it easy for customers to order and buy direct. But companies who face difficult channel conflicts with existing retail distributors could do well to consider this strategic option on its own merits. They should consider how much additional business they can generate through an excellent implementation of an 'information only' response to ES.

### 6. 'Forming a Club'

Consumers in Japan are getting together to buy in bulk from retailers. They are specifying the source, quality, ingredients and delivery method of certain goods and agreeing to buy a fixed quantity. According to a *Financial Times* report:

> 'as these consumer groups have become more organised so they are beginning to use the internet to shop around for the best bargains and are getting more comfortable with the idea of buying electronically.'

For the first time the Internet offers a meaningful opportunity for consumers to take control for themselves of the shopping experience.

Without the physical constraints of location access and hours of opening and with developing software search engines that can shop the Internet to compare prices and value, consumers now more than ever have the opportunity to decide where and when and how to shop – at their convenience.

This poses a challenge to retailers, of course, but also provides a new route to market for manufacturers who can now look up these groups of consumers – these communities of like-minded souls – and build their own direct links with them.

While this development is still in its early stages there are already a few clubs forming and manufacturers have the opportunity both to sell direct to them but also encourage and facilitate their growth:

- One of the fastest growing 'communities' is *SeniorNet* – which has around 20,000 members. It's a non profit organisation whose aim is to build a community of computer-using seniors. SeniorNet maintains forums on America On-line and on Microsoft Network, for example. They cover a range of interests such as the Christian Corner, Divorced Pals and Senior Entrepreneurs. Among other activities the community has started more than 80 SeniorNet learning centres around the US. SeniorNet is now investigating what other services it can provide for its subscriber members. These include combining the purchases made to set up direct purchasing and distribution links with hardware and software suppliers. For such suppliers, SeniorNet provides a targeted electronic sales opportunity. Because the whole operation is still immature there is an open opportunity for various manufacturers to come in as sponsors and supporters and gain a marketing and sales advantage with this growing community of like-minded souls.
- *Motley Fool* is another community of interest web site. This one is targeted at people interested in personal financial investments. The groups of people drawn to this site are diverse but it provides a unique ability to connect people who share relatively narrowly defined interests. The entrepreneurs behind Motley Fool have already begun to sell books and other products related to financial investment and are establishing their own direct lines with stock brokers. For financial service providers, this consumer grouping represents a new opportunity to reach target customers, bypass traditional channels and work out how to establish a meaningful sales relationship directly.
- Procter & Gamble has a long history of building a consumer franchise in the diaper market. For many years it has directly sponsored pre- and post-natal services in certain communities and provides a 'goody-

bag' of product samples to new mothers. It has also held forums of mothers and babies to discuss product usage, baby care and future product requirements. A natural evolution for P&G in this area is to sponsor mother and baby clubs on the Internet, providing the financing infrastructure, giving out information and advice but also setting up for itself direct targeted consumer purchasing groups. Such 'mother and baby' clubs have a natural community of common interest and combined with local home delivery services could provide companies like P&G with a substantial direct business.

## 7. 'Treat as Another Channel'

For those manufacturing companies who are exploring electronic commerce, a number have already made the decision to move beyond an 'information only' path. They are determined to explore the new routes to market and exploit additional target sales and marketing opportunities.

Most of the manufacturing ES activity to date, however, has been in the business-to-business arena. For example, Cisco Systems conducts nearly 40% of its sales – in 1997 c. $10 million each day – over the Internet. And Cisco expects volume sales over the Internet to grow substantially in the next year. Boeing, the aerospace group, has reported conducting millions of dollars of trade over the Internet every day. GE, IBM and Seagate are but a few other examples of companies who have made rapid strides exploiting Internet technology and building high levels of sales. The Forrester Research group predicts the value of goods and services traded business-to-business will grow dramatically and reach more than $300 billion by 2002: 'The effect on business of this new channel of distribution will be unprecedented, spawning dynamic new trading relationships and selling routes.'

This is all very encouraging at the business-to-business level but it appears to have been easier for manufacturers to move more aggressively into electronic selling in that arena. Outside of the consumer goods environment, the balance of power is typically less polarised in favour of one party, there's somewhat more flexibility to sell through different channels. Furthermore, in many business-to-business situations, product technology or branding do still provide a clear pull advantage, relationships generally with customers tend to be less antagonistic and there appears to be greater willingness to work together. Customers tend not to produce their own private-label rival manufactured products!

However, many manufacturers operating electronically and successfully business-to-business, like IBM and GE and others, also have a vibrant and sizeable consumer-selling operation. They will surely be asking themselves why can't they achieve the same direct success with *consumer* end-users. Whatever channel conflict blocks exist, they will surely take the ES and Manufacturer Tests and quickly realise they do have the capabilities to start developing the Internet as a consumer sales channel as well. One good example of such manufacturer initiative in the consumer arena comes from Sony, who are slowly building up their direct selling operation.

- Sony has established its own on-line retail store and it's set up to provide a total entertainment experience as well as a sales opportunity. Its 'station' site is like an on-line urban theme park. Its opening page looks like a neon-lit city street and the neon signs point to games, chat rooms as well as music stations. It's a site that's designed to be explored. The shopping sign leads not only to music sales but to a 'Men in Black' clothing store, a T-shirts counter, videos, books, cinema tickets and inevitably its own hi-fi equipment.

  On the shopping pages every item is described in some detail. At the foot of the page is a quantity box, price and the option to add it to a shopping cart. At present, the Sony site is primarily geared to young adults and older teenagers but there are plans afoot to establish a separate station targeted at older, more affluent consumers groups.

  To date sales are not yet significant (Sony isn't actively marketing the opportunity) but in the meantime Sony is also learning hard about how to make its web site attractive and shopping friendly. Once it gets that sufficiently developed it will be in a prime position to take advantage of faster easier Internet access as it becomes more widely available.

For many companies like Sony, selling on-line in the retail environment is still treated as exploring a new channel of business. For these manufacturing groups their caution is driven by a number of factors, but especially the desire to protect existing retail distribution business. Many will wait for new user-friendly Internet access technology to fuel consumer interest and make volume sales targets more readily achievable. In the meantime, organisations like Unilever, as we have seen, have publicly declared their interest to push down this path, while other leading manufacturers like Procter & Gamble, Rubbermaid, Nestlé and Kraft are getting on exploring how to eventually take advantage of this new opportunity.

## 8. 'Set Up as a Separate Business'

There are even fewer examples of manufacturer suppliers to the retail sector who have already so grasped the ES opportunity that they have established an electronic commerce business unit, with its own funds and infrastructure to operate alongside but also potentially in competition to existing retail distribution-lines.

The best example to date comes from the financial services sector and again from one of its leading suppliers – Direct Line. As has been well-documented, the Direct Line operation has been an enormous success but it took that bold vision to set it up and give it its freedom. Most other insurers have been less ambitious and have tried to keep a tight rein on their direct activities to minimalise the cannibalisation risk. This conservatism has in fact provided an opening for greenfield new entrants, with no existing retail business to protect, to additionally enter this market and take share.

The financial services sector looks the most likely to generate more self-funded, fully resourced, stand-alone businesses selling into a consumer retail environment (see Appendix 2, p. 254). One reason is that retail distributors are typically less aggressive in this sector, allowing provider suppliers a greater sense of operating freedom and experimentation. We now wait to see who will first emerge out of the consumer goods pack, burst out of their web site chrysalis and develop something more substantive in sales terms.

## 9. 'Band Together'

Those manufacturers determined to be among the more ambitious may consider this option. It involves groups of manufacturers with complementary product ranges banding together to provide an alternative one stop shop. Examples might include:

- **'The Coffee Shop'** – a virtual site potentially sponsored by Nestlé including a coffee and tea products shop, a Jakobs Suchard chocolate and confectionery shop, United Biscuits McVities biscuits, Coca-Cola as a soft drinks alternative, Frito Lay potato chips and snacks and Time-Warner magazines and books.
- **'The Household Products Shop'** – a virtual site featuring Procter & Gamble home cleaning products, Rubbermaid sinkware and storage products, Home Depot kitchen and bathroom accessories, Bath Bed and Beyond towels and linen, Toys R Us, and Sears for white and brown goods.

These companies could combine to establish and host web sites or TV home-shopping channels or even more traditional catalog phone/fax operations. Provided they communicate effectively and build awareness and branding they could become the new 'retailers' of the twenty-first century. They could displace the physical retail distributors stuck with expensive real estate and struggling to transition themselves into the new virtual environment. These manufacturers could enter the distribution business greenfield, ally with distribution and warehousing partners to access new skills and competencies and establish their own direct consumer-sales operation. In this way, they could get out from the historic tension and difficulties of dealing with their often powerful and sophisticated retail customers, bypass those 'old-fashioned' ways of distribution and enter the competitive environment entirely shaping their *own* sales development.

**10. 'Go Fully Direct'**

Who's going to be the first to go the whole way, transition deliberately out of existing retail distributors and fully embrace electronic communication and selling? Will the financial services sector be the first arena where provider suppliers ultimately decide that running physical and virtual distribution networks in parallel makes no sense and go the 'full monty' after the lower-cost more profitable electronic transactions?

In retail banking services, the traditional branch banking has already been whittled away by ATMs, debit and credit cards. We already have the 'no-people' branch where the shell that was the fully staffed bank has been turned into a room full of machines. Cash can be withdrawn or paid in (acknowledgements sent by e-mail, mail or fax), foreign currency purchased and information provided on statements and balances. Only the more complex of the retail banking transactions (e.g. loans and mortgages) may still require a human interface. But even with these there are alternative electronic purchase channels available.

None of the retail banks operating today will publicly acknowledge that the physical branch service is dying, but they have all been gradually consolidating and reducing the number of branches available. A few may survive where they play an important commercial development role but stand-alone staffed branches in every town are surely on their way out. Eventually some currently physical banks will bite the bullet.

\* \* \*

**Figure 13.3** Manufacturers' Strategic Options Comparison

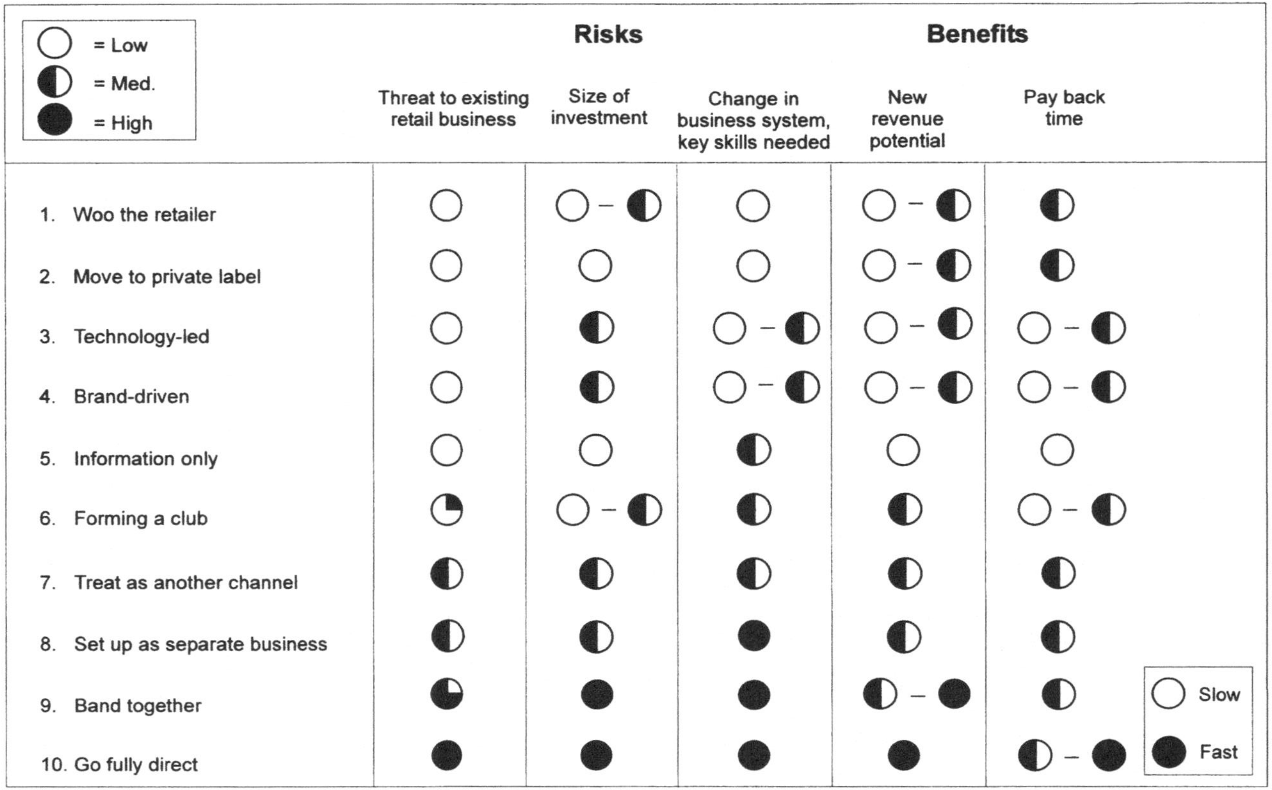

|  | **Risks** | | | **Benefits** | |
|---|---|---|---|---|---|
|  | Threat to existing retail business | Size of investment | Change in business system, key skills needed | New revenue potential | Pay back time |
| 1. Woo the retailer | Low | Low/Med | Low | Low/High | Med/High |
| 2. Move to private label | Low | Low | Low | Low/High | Med/High |
| 3. Technology-led | Low | High | Low/Med | Low/High | Low/Med |
| 4. Brand-driven | Low | High | Low/Med | Low/High | Low/Med |
| 5. Information only | Low | Low | Med | Low | Low |
| 6. Forming a club | Low/Med | Low/Med | Med | Med | Low/Med |
| 7. Treat as another channel | Med | High | Med | Med | High |
| 8. Set up as separate business | Med | High | High | Med | High |
| 9. Band together | Med/High | High | High | Med/High | High |
| 10. Go fully direct | High | High | High | High | Low/High (Fast) |

Legend: ○ = Low, ◐ = Med., ● = High; ○ Slow, ● Fast

201

## Summary

We can now draw these ten different options together and summarise the risk and benefits attached to each (Figure 13.3).

The Partnership options 'woo the retailer' and move to 'private label' naturally carry the least risk in terms of jeopardising existing retail business. In contrast the more direct strategies do confront the channel conflicts head-on, carry greater risk but also potentially greater long-term reward. Manufacturing companies that score high on their test have the flexibility to examine the market place and choose which path they want, how aggressive they wish to be, which risk reward profile they are comfortable with. Others, as we have discussed will have much less room to manoeuvre (Figure 13.4).

Whichever the strategy that's chosen, business success can still certainly be achieved. In each case the winning formula will be a combination of choosing the strategy that best meets customer and market needs plus pursuing an implementation path with rigour, determination and a single-minded dedication to making that strategy work. The best strategies won't succeed, of course, without effective implementation and in contrast tasking the workforce to work hard and fast will inevitably be futile if the appropriate direction has not been set.

**Figure 13.4** Strategic Choices from the Identified Ten

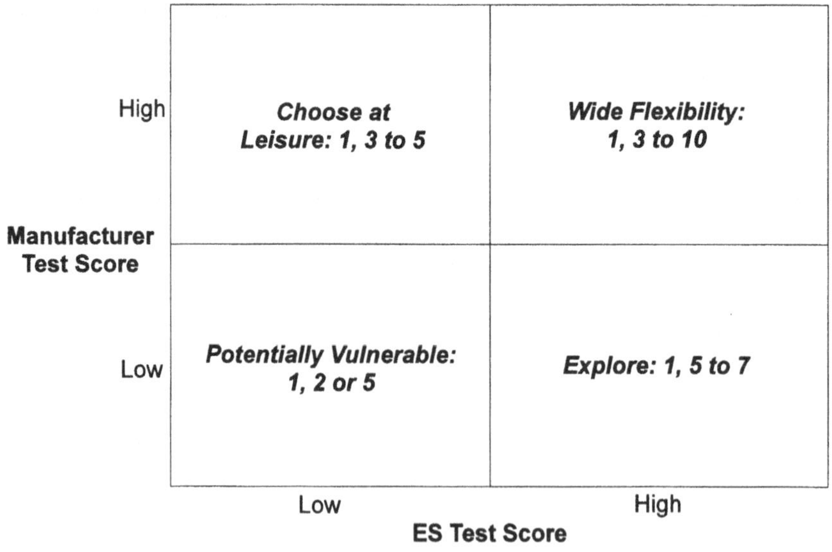

| Manufacturer Test Score | Low ES Test Score | High ES Test Score |
|---|---|---|
| **High** | *Choose at Leisure: 1, 3 to 5* | *Wide Flexibility: 1, 3 to 10* |
| **Low** | *Potentially Vulnerable: 1, 2 or 5* | *Explore: 1, 5 to 7* |

Choosing the 'right strategy' can be guided by the tests and checks we have described. Making the implementation successful can be driven by the discussion and frameworks to be shown in Chapters 14 and 15. But whatever the path, the company must continue to stay close to the market place. It's evolving fast and every organisation must stay on its toes as electronic commerce develops and matures.

# 14 The New Marketing Imperatives: Marketing in the Electronic Age

At least half of multinational companies surveyed are 'missing the link': they're not marketing effectively on the Internet. According to a 1998 study in *The Economist*, many web sites are sorely lacking when it comes to communicating with their audiences. Companies may be on-line but web sites are often difficult to find and difficult to use, information is often patchy or incomplete (only 37% of the companies in the survey gave any details of where their products could be obtained), sales efforts can be half-hearted (e.g. J.C. Penney's handbag department for some time featured precisely one item, the Argos gift service sold only barometers) and on-line support and services can be surprisingly limited (only 21% offered any interactive service or further communication beyond the initial information dump).

There is still a huge gap between the marketing potential of the Internet v. companies' understanding of how to exploit it. But there are a growing number of leading organisations across most retail sectors who are demonstrating that the potential *is* there and showing how it can be realised. Dell, Auto-By-Tel, Amazon, CD Now, Peapod, Tesco, Land's End, Karstadt, Citibank, Directline, NetMarket and many others, in just two or three years from inception, have built impressive electronic selling and marketing operations. They are generating new interest and excitement amongst target customers. They are improving their overall market reputation. They are starting to see significant sales as consumers increasingly get wired up and get used to transacting on-line.

But the majority are still 'missing the links'. They've been in experimental mode since the initial Net frenzy in 1994–5 and still have a long way to go to reach best practice levels (Figure 14.1). According to further 1998 research by Forrester, more than 50% of companies seemed confused about the role of the Internet in their business, unclear what their strategy should be and with no immediate plans to graduate from trial and information to a level of greater selling sophistication. For Forrester, one of the key roadblocks preventing progress is that 'electronic commerce is still the domain of the IT function – marketing departments have not yet gotten sufficiently involved nor claimed any real ownership'. By way of illustration, Forrester calculated that budgets for developing and maintaining an effective web site had risen steeply to around $3 million a year (and sometimes up to $5 million or more). But many spend less than $250,000 and European companies especially fall short in this area.

Can every company make the necessary sales and marketing connections? Can any organisation succeed in electronic commerce? Or does the new medium pose some unique sales and marketing challenge that only a few have cracked? Is it just about applying basic principles or is it inherently more complex? What's the learning that can be taken from the best-practice examples – from the companies already out there reaping the benefits? Can that be translated into a 'how-to' framework of broad application that helps electronic laggards arrive at the best sales and marketing solution? In this chapter we first consider the specific and new marketing challenges posed by the growing electronic revolution and then detail a marketing solutions framework to provide a step-by-step guide to would-be marketers facing up to the new environment.

In developing the marketing framework, the key learning is that organisations must recognise they are indeed dealing with something different. Those companies already successful on the Internet, for example, say that their own break-throughs only came once they acknowledged that the electronic market place does pose new and distinct marketing challenges and opportunities. It's not just a question of

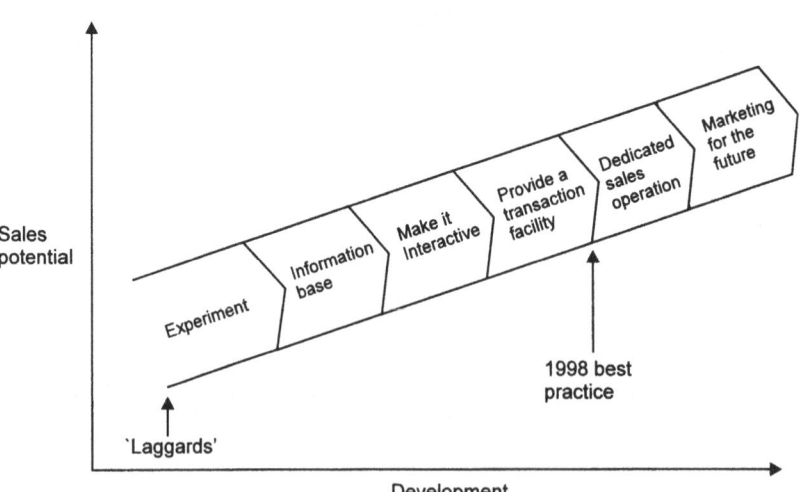

**Figure 14.1** Electronic Marketing Evolution

carrying on as before with an additional channel to market through. With electronic commerce, what can be done and achieved breaks rules and stretches horizons. New approaches, and thus new skills and competences, must be developed.

The particular challenge facing marketers has a number of dimensions to it (Table 14.1). Some have already been referred to and discussed, such as the many new technologies, the number of new electronic connections and the inexorable global reach of the world-wide web. But much of the new marketing opportunity lies in knowledge capture, information management and exploitation. These 'information' areas specifically require further explanation here to fully establish the new marketing scene and context that sets up the solutions framework. The next section therefore looks at the first three of these new marketing challenges in more detail.

**Table 14.1** Seven New Marketing Challenges for an Electronic Age

- More information for sellers about their customers
- More information for buyers
- New intermediaries
- New technology 'interconnectivity'
- More automation
- World-wide sourcing and access
- Greater value consciousness

## Marketing challenges

### *More information for sellers*

In the new electronic environment it will be possible to gather a lot more data on customers. This can be managed to create many new communication and marketing opportunities. For example, Amazon.com is able to track what book reviews customers look up, what categories they are interested in, what they buy, how many and how frequently and where they live. It has a richness of data which it can automatically gather about each and every person who visits its site. From a different perspective, Procter & Gamble can organise virtual focus groups where invited consumers from all over the world can meet in a 'chat room', compare notes, talk about their interests and offer detailed feedback on products. Hotel companies like Marriott and Ritz Carlton, airlines like British Airways and American already have detailed databases on people who use their services, but this can be extended to capture every inquiry, each would-be purchaser who can then be subsequently e-mailed, contacted and incentivised.

The Internet and other forms of electronic connection offer considerable potential for what's called 'database marketing'. There will be better opportunities not just to build profiles of consumers but to track how trends change, look at individual needs and target potential new purchasers. What is new is that the information can be captured immediately on-line and instantaneously processed. *There's a significant competitive advantage for the organisation that thinks through and works out how best to exploit the knowledge that is available.*

Financial Services companies like First Direct and USAA have been in the vanguard of the knowledge bandwagon. Another specific example comes from a relatively new company on the internet called Firefly. It proactively seeks permission from people who visit its site to gather more detailed information about them. (It's been especially keen to reassure about privacy and data confidentiality and uses Coopers & Lybrand to audit its safeguards in that area.) Firefly then builds psychographic profiles of customers and uses that to target other potential customers of a similar disposition. The knowledge it obtains is used to pinpoint carefully tailored advertising messages. Its knowledge about its customers also attracts other advertisers to its site who want to reach the same target audience. So Firefly gets a double benefit from its information capture – improving its own marketing and exploiting its niche to attract others.

Firefly is but one small illustration of how electronic connections can provide a continuous stream of personalised information. If it's captured efficiently that information can provide new insight and understanding and help the development of tailored sales and marketing initiatives.

*More information for buyers*

Just as sellers can learn more about their customers, so customers can get a lot more information about the products and services they're interested in. Their reach through the Internet becomes global, they can sift through vast amounts of often detailed specification that might otherwise not have been easily attainable. They can look at samples. They can review extracts from movies, newspapers, books and music. They can compare prices. They can interact with others sharing a similar interest and get 'inside' feedback on what the product is like. They will be able to experience the product more vividly in three dimensions. And they will be able to do all these things at their leisure, at their own pace and when it suits them.

Buying over the Internet puts the consumer firmly in control of the process. But it also makes them harder to reach and market to. They're not caught up in a retailer's catchment area compelled to shop only where they can drive to and select from only what's on the shelf. They're not limited by the manufacturer's brochure and what information they happen to glean from the salesperson. Nor are they constrained by a manufacturer's distribution and lead times. They'll be in a better position than ever before to call the shots. Bargaining power shifts firmly in their direction.

In the new electronic age, marketing must take on this new dimension. It needs to become even more of a two-way process. Creating awareness, stimulating interest but also being able to respond to that interest, deal with requests, capture passing surfers, manage interactions, develop relationships. To reach out to customers and win their business, sellers are going to have to be much more tuned-in to a dialogue that's interactive. They'll need to embrace all the marketing 'Ps' (place, product, promotion, people, price, packaging) as never before. The marketing framework set out below can be their guide.

*New intermediaries*

As the electronic value chain takes shape, new companies will emerge with dedicated competences in focused areas. They can provide best-practice solutions or help other organisations achieve stronger market positions.

One such opportunity may come in the information-management area. There, new intermediaries will develop who start off providing cost-efficient storage capabilities. But they may evolve into processing and integrating massive amounts of data. They may then progress to providing additional value by sorting and reporting for client companies. At every stage they could sell information or insight to others to improve their marketing databases.

Other intermediaries may appear in the logistics part of the chain. Dedicated warehousing, transportation and home delivery operators could enable companies themselves to go virtual, outsourcing most or all of their previous physical infrastructure. Especially in home delivery, there looks to be a substantial opportunity to establish a local neighbourhood network that builds strong ties with the consumer, ultimately taking responsibility for all things delivered to the home. Such an organisation could end up replacing the retailer as the company closest to the consumer. They could have the best database on consumer habits and needs and through that offer the best value-added and personalised service.

Back up the value chain, Internet access providers like America On-Line (AOL) are gradually evolving their role into powerful intermediaries who can control and influence where consumers go shopping. In 1998 AOL had nearly 20 million customers. Would-be electronic sellers are beginning to realise that they need to be present and have a high profile on AOL if they are to make sure customers can easily find them. AOL has taken on the value-added role of location manager and companies are willing to pay significant sums for prominent placement on high-traffic web sites such as AOL's home page. For example both Amazon and 1-800-FLOWERS recently announced they had entered into long-term exclusive agreements with AOL. Amazon has agreed to pay $19 million and 1-800-FLOWERS $25 million plus a share of revenues. In addition, CUC International is paying AOL $50 million, again plus a share of revenues, for becoming AOL's exclusive discount shopping service.

As AOL Networks president Robert Pittman commented:

> 'We're in the land-grab period and this is the Malibu real estate . . . AOL is going to be of increasing importance to get access to home and at work users . . . we control the access to the customers in the electronic environment and sellers have to be there.'

'It's a land rush' commented the *Wall Street Journal*, 'for what many consider one of the greatest and least developed parcels of territory in cyberspace.'

### The marketing solution framework

Now we understand more about the marketing challenge the 'marketing solution' can be set out. A framework of ten steps can be identified that can act as a guide for successful electronic selling and marketing. It is drawn from research and interviews with those companies like Amazon and Dell who are demonstrating current leading-edge approaches and best practices. It has a mix of basic underlying marketing principles as well as embracing the new competences required to succeed. It should be no surprise, then, to see fundamentals such as 'understand the customer' alongside relatively new ideas like 'interactivity' and 'networking'.

All ten steps must be followed through. The learning is stark: miss out one of the steps or carry it our half-heartedly and it can seriously undermine the total effectiveness of the marketing campaign. Winners

**Figure 14.2** The Information and Knowledge Master

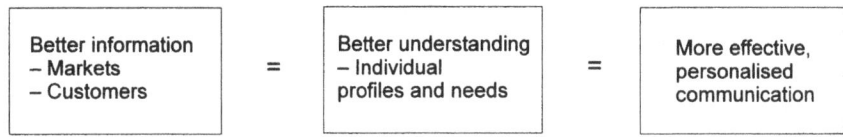

appear to be aggressively pursuing and investing in every part of the framework. But if there's one overriding theme that runs through the framework it can be captured in this way: become an information and knowledge master (Figure 14.2).

The ten steps of the framework focus on effective marketing on the internet, to pick the potentially most powerful medium. The individual steps form a total integrated process or value chain (Figure 14.3).

**Figure 14.3** Marketing Solutions Framework

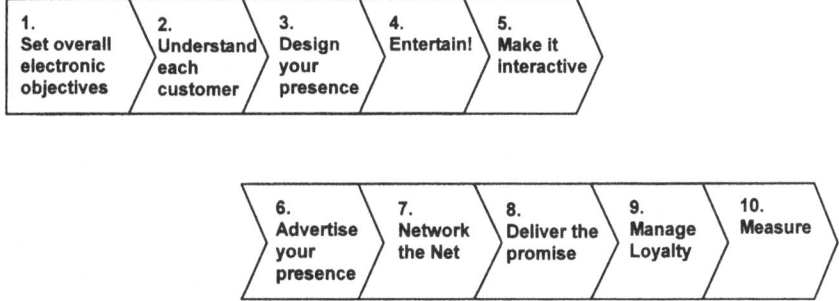

*1. Set overall electronic objectives*

Research by Forrester, the *Wall Street Journal* and others continues to show that many companies are unclear about the role of their web site in particular. Initially many could genuinely claim it was early days and they were researching and working out what sort of web site to develop. But customers' patience with 'web site under development' notices, single handbag offers or obviously incomplete home pages is fast running out. 'Best practice' examples of what a good web site can do are already much in evidence. They are setting standards and expectations. Those that fall short are now especially conspicuous and are beginning to generate negative publicity for the entire business.

Companies have a choice of role on the Internet and in the electronic arena generally. They can develop it either for promotion or for content or for transactions – or a combination of all three. But the choice must be deliberate.

- Promotional sites are about creating awareness and reinforcing brand image and reputation. They deliberately stop short of providing detailed information. But to be effective they need to signpost where that information can be obtained. They also deliberately stop short of providing transaction facilities but to be worthwhile they have to set out where the product or service can be purchased.

  A number of consumer goods' companies like L'Oreal, Nabisco and Kraft featured sites of this type where the key role and emphasis is brand reinforcement and image-building.
- Content sites that are successful are designed to inform, and entertain, with material specifically designed to work well in an Internet environment. Levi's and Sony are two examples of companies who've gone beyond the promotional. They are attempting to attract an audience in their own right with games, features, news and chat rooms. Such sites usually take six months or more to develop and often contain 2000 pages or more of content. They are demanding in terms of resources, requiring a dedicated team of often 10–20 people to update information content on a daily basis and continue to keep the site alive and fresh. There are usually interactive elements with their sites in the form of e-mail or message centres.

  These sites can be exciting places to visit but a number of companies anticipate that their 'content' role is only a stepping stone till they become fully-fledged transaction bases offering the full range of goods and services.
- Transaction sites may also contain plenty of content but their principal objective is to sell product. Forrester estimates that such sites can require staff of 30 plus people to operate and maintain. They will have clear sales targets and no doubt in these early days of the Internet will be expected to generate incremental revenues over and above the base business.

Making a choice of Internet role is critical. It determines the reasons why a company is setting up electronically, what it is trying to achieve. Is it simply another advertising medium or is it a sales channel with specific targets? Is the goal here to entertain or provide a full sales and service operation in its own right? Making clear decisions in this arena will drive investment and resourcing, identify new skills that need to be developed, establish network partners to provide the total Internet package and the communications that must be made to customers. The choice itself of Internet role is unlikely to be at the providers' unfettered discretion. Taking the ES Test will provide clear guidelines on consumer interest levels in the medium generally and in purchasing in particular. More detailed research will show whether customers are interested in the whole product range or in some parts only and more precisely what type of promotion, content and transaction facilities they require.

## 2. Understand each customer

For companies that are already market and customer-driven, this step in the process is straightforward and plays to core skills and understanding. Such companies will have been among the first to take the ES Test, they will have further developed their customer segmentation beyond the six basic categories laid out in that test (see Figure 5.8), they will have their own tailored and targeted insights attempting to understand each and every customer.

Database marketing and the information-rich environment of the electronic age allows this level of detailed understanding to be distilled and processed and still be available in convenient intelligible form even though the customer base may number millions of individuals. Examples from the insurance industry have provided early illustrations of what can be achieved. USAA has such a sophisticated database and such a developed call-centre environment that when people call in the telephone operatives have on their screen details that allow an intensely personalised call: 'when we last spoke I see you bought life assurance, are you satisfied with that still? I see it was your wife's birthday last week, how are your two children, you may be interested in our new family health care service that we have just set up . . .'

Fred Meyer, the regional grocery chain, provides an intelligent approach to understanding the customer in an electronic age:

'We understand that in the grocery industry the internet must play a role in our future, we know that a significant number of our customers expect this and we are experimenting with it at various levels. One area where we've learnt there is particular interest in internet shopping is among our customers in Alaska. Most of the Alaskan shoppers travel hundreds of miles, some of them live days away and many can only come in the spring and summer. So we started offering a service over the internet that enables them to do their shopping. We take the orders into one of our seven stores in the region where they pick the items but then we have to get them out to the customer and normally that means by bush pilot. Now the cost of doing that will make most say just forget it. But we are looking at supplies that normally last from three to six months. We are experimenting with this in the Alaskan market, adding on various complementary products and services and seeing just how much business this can amount to.'

Understanding the customer is a basic principle for doing business. There is no substitute for the Triple I model (Figure 5.6) and approach. The Internet environment makes this market immersion, intuition and innovation process all the more challenging but equally more rewarding for it can provide the more personal information detail that enables the individual customer to be targeted even more precisely with what they want.

### 3. Design your presence

Research among Internet surfers and customers throws up some satisfied shoppers but many frustrations: 'can't find my way around, too much information, not enough information, too many distractions on the page, too many steps between access and ordering' (16 clicks is not uncommon versus a target benchmark of 3!). Sites are often poorly laid out and suffer too from trying to cater for different types of consumers – some only who want information, others who want to play one of the games, still others who want to look through a range of products to choose what to buy and some who know precisely what they want and just want to go straight to the order page. No matter who they are, what their interests, consumers still typically have to go through a standard page sequence and they have to put up with the same level of often superficial language whether they're a teenage surfer or a corporate finance director.

Looking at best-practice sites today such as Dixons (the UK electricals retailer), Auto-By-Tel, Karstadt's 'My World', AT&T, World Net and Dell, these sites are laid out and designed according to some simple rules that research can identify:

- the purpose of them is clear – if they're intended to sell product then they provide clear order – pay sequences
- there is an overall site map for navigation

- it's targeted at distinct types of consumers, e.g. the teenage surfer is immediately directed to a different site via a click-on icon allowing the would-be shopper to get on with the transaction
- information and news is dated so the customer knows how current it is
- once a satisfactory process is established, stick with it and try to resist constantly changing it, evolve it within certain parameters
- provide click-on access to complementary products and services.

## 4. Entertain!

An Internet site can provide a wealth of experiences. There is the unique opportunity to build images, video and text and do so in a creative way that captures imagination and interest. Beside the simple convenience that can be provided by electronic shopping there are many other reasons consumers can be persuaded to visit the site or spend more time there associating with the brand and the products and naturally building a closer affinity to them.

Successful sites do more than just provide content or transaction facilities. They involve and entertain. They engage consumers in a way that teaches them more about the product/service and why it is better than competition. They provide this learning in a fun and subtle fashion and by providing different incentives encourage repeat visits – to play and, if appropriate, with the option to buy, always just an icon-click away.

Procter & Gamble has a site for its Tide Laundry brand. Apart from some basic information, P&G invites browsers to play the Tide 'Stain Detective' game: 'If you spill red wine on your favourite cotton shirt you can ask the Tide detective's help. He will help you find an answer that is fast and proven.' American Airlines holds a daily airline seat auction that invites consumers to bid for spare seats in up-coming flights and wait to see if they've won. General Mills has a created a top children's site called You Rule School that's chock-full of games, education materials, news and features on top bands and new movies. Gap has a 'try-on' sequence that shows how different outfits can look and a 'mix and match' section that shows which accessories could complement the preferred choice of clothiers. Gap is working hard with software houses to get 3-D displays that could significantly enhance these 'try-on' experiences.

Other organisations ranging from Bristol Myers and General Motors through to start-ups like Kidsoft have begun to appreciate the entertainment opportunity, and in doing so are setting standards for others. It's this experience factor that differentiates the Internet even more starkly from most current shopping environments. It's the added draw that brings in new users as well as sustains the interest of existing and frequent customers. When appropriately designed, it's reinforcing of the total product proposition and is a key ingredient for making Internet sales incremental to the total business base.

## 5. Make it interactive

There's been a lot of hype about interactivity and how the Internet creates a unique environment for it. But it must be said that any effective marketing campaign – whether it's on the Internet or through more 'prosaic' media such as print or TV – should always contain some interactive element. Successful sellers will always be trying to engage their customers, get down to individual requirements, have some conversation with them, answer questions and move the prospective buyer towards a purchase decision.

Certainly the Internet provides more opportunity to do that and to accomplish it in a more sophisticated way. It can provide something close to a genuine dialogue in real time between two parties, 'listening' to customers' queries and providing an immediate response. The particular attraction for consumers is they can manage the pace of this interaction. They can get what information they want when they want it. They can make several visits to the site without feeling under pressure to buy. They can take their time making their minds up. It's an attractive interaction for sellers, too, because as they're indulging in the consumer's questioning so they can be learning and capturing data about the type of person they are dealing with, what are their points of interest, what their hot buttons and how they can best be responded to. A sophisticated Internet interaction can be as powerful as fielding your best salesperson, listening, learning and adapting the 'sell' to meet the customer's needs.

One example of this is a US law firm based in California that used to send out an annual and rather weighty booklet called *Intellectual Law*. Subsequent research showed it mostly got sent for filing in clients' libraries or gathered dust on shelves. The lawyers saw an opportunity to use the electronic medium to change the whole communication context and get their clients interacting with the material. It was first established as a CD-Rom and then subsequently put on the Internet itself. It was designed to be a marketing rather than simply an information-giving tool and prospective clients were asked to take a quiz – an entertaining ten-question test to identify what gaps if any there were in their knowledge in this field. There was an interactive e-mail facility and a daily schedule set out when individual legal experts would be available to field particular questions. It was set up so that some basic issues could get resolved in that way and it was possible for potential customers to dig deep into the subject. The basic interactive service was free. The CD-Rom version was installed in kiosks and displayed in the lobby area of the law firm's offices. The firm says this electronic service has definitely enhanced its reputation as a leader in its field. It has new clients who say it was this innovative service that brought them into the firm's net.

### 6. Advertise your presence

Companies are beginning to learn about the power of advertising on the world-wide web. In 1997, advertising spending was forecast to hit $1 bn by the year-end – that's a near eight fold increase vs. the previous year and it's expected to carry on growing at very fast rates. 'Net-advertising', according to Andrew Grove, CEO of Intel, 'is becoming a big deal' and many companies are ramping up their web advertising plans:

- Bristol Myers Squibb teamed up with software-maker Intuit Inc through 1997 to launch an on-line advertising campaign for Excedrin one of its over the counter (OTC) medicines. For 30 days Bristol Myers ran ads on financial web sites offering free samples to net surfers who clicked on the ad and typed in their name and address. Just during that trial Bristol Myers added 30,000 new names to its customer list and that was three times the company's best estimates. What's more, the cost of obtaining those names was substantially less than other traditional marketing methods. The ad agency that produced the campaign commented: 'We've turned a packaged goods company into QVC.'
- IBM has announced that for 1998 it will place ads on 500 web sites increasing its advertising spend on the web threefold. Microsoft has said its spend will top $30 million.
- E*Trade Group Inc. has been an early success on the Net. This financial services company specialises in on-line stockbroking and has established an integrated advertising campaign using print, TV and

the web itself. E*Trade started off spending c. 5% of its media budget on the web but now expects that to rise to 15–20%. Its early 1998 campaign on Yahoo!'s finance site brought 3118 new customer leads in ten days. As the company's CEO has said: 'we can't imagine growing our business without this, a media strategy without the Internet would be suicidal'.
- Toyota had 7 mn web surfers visit Toyota. com in 1997. More interactions than on its 800 number. Toyota has now signed up Saatchi & Saatchi to produce a new Net advertising program integrated with its print and TV campaigns. Toyota found that as a result of its advertising c. 170,000 people typed in their names and addresses and over 7000 went on to buy cars – a remarkable near 5% conversion rate.

Advertising interest was originally dominated by the computer companies. But there's been an explosion of interest from makers of everything from Toyota cars through to Kellogg's Corn Pops. Excite Inc. the No 2 web search company, saw its proportion of non-tech advertisers rise from 38% to 59% in the last six months of 1997. General Motors, American Express, Walt Disney and Procter & Gamble have all become relatively significant spenders and for these companies and others it's becoming part of their mainstream advertising planning, like TV or print.

A recent innovation is known as the 'interstitial' (soon no doubt to find a more colloquial label). This is a full screen that pops up either in the lag time between requesting a web page and its appearance on screen or between segments of information. For example, one site featuring the quiz show 'You Don't know Jack' has attracted such major advertisers as Seven Up, Hugo Boss and 20th Century Fox who reach out to the site's 150,000 players each month showing 'interstitial' ads of up to 15 seconds including previews of forthcoming movies. The cost is less than $20,000 which is a fraction of what companies normally pay to reach such customised target consumer segments.

And Internet audiences are fast becoming sizeable enough to provide significant contact and communication to a number of target sectors (Table 14.2). This makes cost per thousand (CPM) comparisons with traditional media begin to look favourable. Versus direct marketing costs, estimates suggest that web advertising CPM on a well targeted site could be as low as one-fifth. Versus a mass TV audience the Internet still is more costly as it hasn't yet achieved that mass penetration but the differential is reducing all the time as more and more households get wired up and connected.

**Table 14.2** Internet Penetration of Specific Consumer Segments

| Target group | % of total segment size using the Internet in past 30 days |
|---|---|
| Students and young earners (18–24) | 20 |
| Young women (25–34) earning more than $20,000 | 21 |
| Young men (25–34) earning more than $40,000 | 26 |
| Men (35–44) earning more than $75,000 | 43 |

*Source*: Simmons database, McKinsey analysis, 1997.

Saatchi & Saatchi have found through 1997 and early 1998 that the vast majority of their clients are now turning to them asking how they can use the Internet as an ad vehicle. Saatchi's clients include a large number of the world's top 100 advertisers and so the significance of their awakening interest cannot be underestimated.

## 7. Network the Net

Location, Location, Location: if consumers don't know about the site or cannot find it then their custom is lost. Just as this is the fundamental rule for all physical retailing so it just as critically translates into the Internet environment. If anything, it becomes an even more fundamental challenge because there's no neighbourhood presence or catchment area that automatically brings a seller to a customer's attention. It is a virtual world and companies are going to have to work harder to brand, build and draw attention to themselves.

One route being especially developed is to locate ads on others' sites where there is some related or complementary product or service being described. The intent is to scatter a series of signposts and billboards across the information highways to direct traffic to your rest stop:

- **Banner ads** are usually small rectangular graphics that sit like roadside bill boards typically at the top of a web page. They're usually static, though new technologies are looking to make them more animated
- **Button ads** are small squarish ads that are placed at the bottom of a web page, often only referring to a brand name. Clicking on the button takes consumers directly to that brand's web site – for example, Netscape have their button ad on most web pages and a click on it takes users instantly to the Netscape web site to download the Navigator browser software.

- **Sponsorship or co-branded ads** attempt to get information seekers on one product to associate that with another. So Toys R Us sponsors a toy guide on the Third Age senior citizens' web site, showing the older generation the toys their grandchildren are likely to be asking for.
- **Keyword ads** feature primarily on web search engine sites such as Yahoo! So Miller Brewing has 'bought' the word 'beer' on Yahoo! and every time someone conducts a search using that term, an ad for 'Miller Genuine Draft Beer' will appear.

Companies will also have to make much greater efforts to integrate their marketing across the different media available. They will need to determine the role of each medium they choose to advertise or sell through and work out how each can complement the other. There are a range of choices now available, of course, ranging from traditional physical communication via direct mail, retail outlet or salespersons through to the electronic media of 800 numbers, direct phone lines, direct PC links, fax and the internet. Should a company run both 800 number call centres and Internet sites for example? Should each connection contain details of the other? Does the Internet address appear on all letter heading, packaging and other advertising? Or should there be a segmentation of channels and communications directing target consumers to a more limited selection of communication options right for them?

The 'networking the net' challenge has been put most succinctly by Doug Cormany, president of the IBM systems' users' group and VP at retailer Fred Meyer:

'we all know how important adjacencies are in terms of store layout and department connections in a physical store. The same thing can be applied to electronic retailing. And it's even more powerful because we have the opportunity for hot links, not only to complementary retailers but to manufacturers as well. We need to think about merchandising this electronic business in the same way we think about merchandising the physical store today.'

## 8. Deliver the promise

After all the hype and excitement of entertainment and banner ads and networking the site, it's clearly going to be important that the product or service does get delivered on time and in full and to the satisfaction of the consumer. Otherwise, no matter how attractive the site, if the logistics don't fulfil their side of the bargain then the customer won't come back and word-of-mouth recommendation will not spread in the way it was hoped.

As was discussed in Chapter 6, home delivery logistics is likely to be a new competence and skill for many companies. As they learn and get up the experience curve they're likely to make any number of mistakes. Only a few companies happen to have these logistics already established and in place and this is likely to drive substantial competitive advantage in the short term, at least till others catch up.

One such example is Sears Home Services. This division of Sears US turned over some $3bn of revenue in 1997 but expects to triple in size very quickly as demand for home services generally grows. Importantly, it has the network in place. Its principal business has been home repairs and maintenance and has an established workforce of some 15000 operatives who have been professionally trained and who Sears claim have an average 15 years' experience. They visit millions of America's homes every year, reckoning they cover some 20% of all households. With a sophisticated communications network already in place, isn't Sears a significant step ahead of most other retailers and manufacturers in being able to deliver to the consumer? They may be sitting on a goldmine of an operation – which could become a home-services/home delivery powerhouse in the emerging electronic age.

## 9. Manage loyalty

In this increasingly competitive and complex world it's not enough to think about customers in the context of a one-off transaction. A business cannot afford to lose customers at all, nor can it just serve them once and not care less whether they come back. From the moment contact is made, companies need to be thinking about a relationship, a connection that can last a lifetime. The art of keeping customers is akin to a marriage, the cost of losing them now the equivalent of a divorce.

The challenge is made even tougher in a virtual world where freedom to choose is so much greater, where the consumer can access every available supplier across the globe and on each purchase occasion readily compare prices and other features and make a most informed value decision. It's easy for customers to be promiscuous and they often perceive no disadvantages or costs to them in switching suppliers. Indeed the research shows that on average most companies will experience defection rates of 10–30 per cent a year.

Can this customer instinct to check out different suppliers be in any way contained? Can companies do anything material to develop relationships with customers that can be lasting and satisfying on both sides? The answer lies in developing a better understanding about what drives customer loyalty and what incentives and levels of performance need to be delivered to keep the customer totally satisfied.

There is no question about the value of doing this. Numerous studies show that retaining a customer pays back many times. Reichheld and Sasser, for example, in their research showed that a 5 per cent increase in loyalty could lead to a 25 per cent increase in profitability. Among other things, expensive 'acquisition costs' for each new customer can be avoided and the studies also show that customers who keep coming back typically spend increasing amounts with the same supplier.

But achieving these loyalty gains is hard. It requires an extraordinary level of effort, motivation and commitment throughout the company that gets each and every employee determined to make a personal contribution to the loyalty effect. Companies like Xerox, Ritz Carlton Hotels and others have won Baldridge quality awards for their high levels of customer satisfaction and for the quality of product and service made available right the way through from the first contact with the customer to the last.

Providing incentives or specific loyalty-based schemes has become popular in recent years as a way of topping up individual service activities and providing additional incentives to keep people coming back. The best known schemes, of course, are the air miles programmes developed by airlines and these have been copied by hotels, credit cards and more recently supermarkets. Sainsbury, for example, launched its Reward Loyalty scheme in 1996 and within a few months had 7 mn cardholders. That's one-third of all UK households signed up with a specific bonus and incentive plan to keep visiting and spending more. Another illustration comes from the Dutch food retailer Edah which has established a smart card interactive Kiosk operation in all its stores designed to build loyalty. Shoppers can apply for smart cards with personal details about themselves and their shopping preferences on them. They insert their card at the kiosks in the store. They are greeted by name and immediately referred to special offers targeted at them and based on what they have purchased previously. Take-up of the smart card system has been high and the touch-screen system also has novelty quick games and questions which generate target coupons. Edah reckon this loyalty scheme has by itself contributed to a c.5% growth in like-for-like sales.

The flexibility and sophistication of the Internet can help companies set new standards in loyalty-building and in 'relationship marketing'. They can achieve a depth and breadth in the information they gather that gives greater insights into what are customers' needs not just for this one transaction but on a longer-term basis. Customers do appear responsive to specific loyalty building programmes and at relatively little cost it is possible to dramatically improve retention rates – with all the profit upsides that brings. For marketers keen to take on the new electronic age and find new forms to build the business this will become an increasingly critical area to innovate and master.

## 10. Measure

Marketers want to be able to measure the effectiveness of what they do and how they spend their money. TV advertising, for example, has developed sophisticated measures that record among other things how many people saw an ad, at what time, what type of demographic profile those people represent, what their typical weekly shopping basket contains and what are their preferences. The challenge now is to translate a degree of that measuring sophistication into Internet marketing activity.

Already a number of measures have been identified:

- **Cost per thousand**: a standard measure to compare advertising costs of reaching target consumers
- **Cost per click**: a new way of evaluating advertising based on the Internet surfer clicking on response to a displayed ad
- **Click through**: measures how often a viewer will respond to an ad by clicking on it
- **Cost per head**: measures only those people who respond to an ad with personalised information such as an e-mail address or other details about themselves enabling the company to follow-up
- **Number of hits**: measures the number of files retrieved.; accessing one single page containing hyperlinks to, say, ten other files would count as ten hits
- **Number of visits**: an approximate estimate of the number of individual surfers to visit a site; it is usually a derived number dividing the number of hits by a factor representing the average number of links per page.

Advertisers are becoming increasingly concerned at their ability to report on how well their marketing investments on the Net are doing, how they compare with alternative media and whether they are paying back. In a recent seminal speech by Bob Wehling, a Procter & Gamble Vice President, there was a call for the advertising industry to get its act together:

> 'we need to establish clear, broadly-accepted standards for measuring cost effectiveness. We have no doubt the Web can be a highly cost effective marketing medium but to become truly valuable it must demonstrate it is highly efficient. We need accurate and reliable measure to guage that efficiency. The measurement systems today are nowhere near accurate and reliable enough.'

One consequence of this is that P&G has so far adopted a conservative approach to web advertising and has insisted that it will pay for ads only when someone clicks on them. As a marketing heavyweight, P&G typically sets standards that others eventually follow. But without alternative measures, P&G is in practice forcing service providers and search engines to share the risks of this new medium. As Yahoo's CEO Tim Koogle commented: 'it's like someone saying you're going to deliver the ads and you're going to have a vested interest in increasing the performance of them'.

The latest initiative in this arena has seen accounting giants Ernst & Young, Coopers & Lybrand and Price Waterhouse setting up specialist units to audit claims made by web sites about their audiences and the numbers of visitors and hits a site says it gets. For example, from the beginning of 1998 Microsoft started sending to all its advertisers quarterly reports of all its web sites audited by Coopers & Lybrand.

As Percy Barnevik, CEO of ASEA Brown Boveri has commented: 'only things that are measured actually get done or improved upon'. Without recognised and reliable measurement systems marketers will struggle to convince their Finance department colleagues to invest more and truly test out and stretch the potential of Internet selling and communication. The new medium is still in its infancy. It took the TV Broadcasting industry and associated companies fifteen years or more to put reliable measuring mechanisms in place. With the internet and the general explosion of interest and revenues, the demand and need for an adequate measurement environment will undoubtedly ensure that some effective framework is put in place much sooner.

\* \* \*

Let's leave the final words on this marketing challenge to Bob Wehling, the Procter & Gamble VP quoted earlier:

'We're moving down the information highway at warp speed. Our destination is the future, the future of marketing, actually a future where smart marketers will be surfing across the Internet into the homes of billions of consumers in hundreds of countries, at the speed of light. We'll be driving by the virtual storefronts of other companies who are also participating in and creating this future. Pay attention because we're all going to have to get wired into this . . . The key marketing challenge is to tap the *full* potential of these exciting new electronic media especially the Internet. How do we make it a global opportunity for advertising and a rich source of entertainment, information and community for consumers . . . what's the add-on sales potential? What's important is to remember it's not just about piping a 30-second ad over the Internet or converting printed materials into electronic form. It's going to require a fundamentally different approach to succeed.'

# 15 Setting the Strategy and Mobilising the Organisation

78% of senior executives in a recent study felt that electronic commerce would significantly change their organisations over the next ten years. They anticipated there would need to be radically different structures and to be effective they would have to develop new competences and skills and manage an inevitable cultural upheaval. Similar findings came through in a detailed study by *The Economist* involving more than 400 business leaders across 34 countries. Two out of three interviewed said that their organisation in 2010 would be very different from today and that there would need to be a keen emphasis on flexibility and faster response. Forrester Research, who have been closely monitoring the progress of electronic commerce, believe the new strategies required will cause 'massive disruption and will challenge many corporations' basic operating model'.

There seems to be growing recognition of the impact that electronic commerce will have – not just in the market place with customers – but on the way the whole company gets structured. It's a business challenge that's going to be so significant that it cannot be left to one function to sort out. It's not just a matter of some new software for the IT department or more funds for Marketing. It's becoming an organisation-wide phenomenon requiring an organisation-wide response. It's potentially sizeable enough in revenue terms that large parts of existing physical distribution structures may either no longer be required or may need to change radically. And it comes on top of a competitive environment that's already fast-changing, becoming more global, consolidating, deregulating and demanding, with customers searching aggressively for greater innovation, more comprehensive service solutions and better value.

What are the most specific organisation threats and changes, what structural, skill and systems issues will managers have to deal with? Will the whole organisation truly be affected or can any restructuring be contained to one or two areas? And once the impact on a company has been identified, what will be the keys to the most effective response, what process or actions will help the business transition efficiently to the new operating models and ways of building business? What role should strategy-making play in determining how best to move forward? How can senior managers make sure their whole workforce is engaged and mobilised to drive the new changes and structures through?

To address these questions we should first understand what are the specific organisation impacts that companies will have to deal with. We can then review what are the key success factors in strategy-setting that can help shape and determine the type of structural responses required and then how to engineer the most effective people-empowered implementation.

**Organisation impacts of electronic commerce**

Reviewing company experiences of electronic commerce and work done especially by Don Tapscott, President of the New Paradigm Learning Corporation, seven issues commonly get cited in research and interviews:

- Digitisation is removing the need for human interface
- Virtualisation renders physical location and proximity irrelevant
- Networking enables everything to be outsourced or done through alliances
- Knowledge capture or 'information flow co-ordination' needs to become a central, driving core competence
- Disintermediation requires new supply lines
- Convergence demands a new relationship between Marketing and Information Technology
- Organisation agility in terms of adaptability to change and fast response becomes the necessary platform on which to build success.

*Digitisation*

Information in any form – pictures, graphs, video and audio – can now of course be digitised and compressed and transmitted fast to any link anywhere in the world. Most parts of a business transaction can be reduced to bits. Exchanges are fast, often instantaneous and error-free. Once equipment has been installed, software customised and debugged, business can take place automatically. There's no need for any human interface. Let the machines communicate! They save time. They save money. They cut out human imperfection. They remove personal prejudice or embarrassment that might have got in the way of a sale. EDI (electronic data interchange) has already been established for some years and a number of companies, like GE, like the major retail chains, now insist their suppliers link with them in this way. Tendering for business, orders, invoices, queries, delivery notes, messages can all be handled electronically and automatically. Once the contract is established, re-ordering can take place with little or no human intervention.

Apocalyptic stores of factories run by robots, offices without paper run by computers have long featured in dramatic visions of our future world. But the sheer speed and capability now of computers makes a people-less, paper-less environment ever closer and more achievable for those who choose to advance down that path. What's emerging, however, is that business generally may have no choice about taking this digital and electronic route. Commerce is moving that way. Competitive pressures, the inexorable demands of stakeholders for growth and added-value, may well force the pace and direction of change for many corporations. If one company doesn't do it, it might get beaten by another taking advantage of the electronic world. Preparing for that is tough. Getting there will involve ever-more wrenching change. But building an effective and powerful IT environment will be the vital platform to provide the flexibility and capability to manage any such transition.

If electronics replaces people then what role can individuals still play in the value chain? That will be one of the major challenges for corporations moving through the next two decades. Dell's response is to see this as a major opportunity to get their employees out there with customers on site assisting in their customer's operation with expertise, know-how and general customer care. Might the new world of electronic commerce act as a general catalyst to move the organisation's focus and the majority of its people at last away from internal administration and instead more generally out into the market place, finding new ways to add-value with the customer base?

*Virtualisation*

This is an inevitable consequence of digitisation. Physical space and distance become irrelevant. The virtual corporation will become the norm. Teams of people will come together, discuss and resolve issues and make plans without ever physically coming face to face. Shopping, chat, information exchange, education and play can all take place in a virtual arena.

Why, then, build more administrative office space, why not invest instead in better computer training and equipment for employees? As has already been pointed out, why build more shops when one option at least is to conduct all business electronically? Why keep printing paper when information can be stored and retrieved in paperless form? Why keep printing money when the future lies in plastic smart cards?

42% of executives in *The Economist* study referred to earlier believed that their companies would evolve to a more or less virtual state by 2010. It conjures up a future scenario which is both intimidating but exciting. Changes along these lines appear inevitable, but no doubt the pace of change will vary by industry sector. It also looks likely that US corporations will confront these issues first and be in the vanguard of working out how best to deal with them. Future winners, no matter what country they are born in, must take a proactive view and get their organisations to understand these changed dynamics. They will need to take a very different view of their asset base. (See further in Chapter 10 on how intangible assets may gradually replace fixed assets.)

*Networking*

As part of the move toward virtualisation, one response many corporations are already considering and testing out is to build value networks. As discussed in Chapter 6, this is a grouping together of different suppliers, partners and outsourced operations to bring together the best skills and integrate them into a cohesive proposition to customers in the market place. It's a recognition that trying to do it all in one corporation may just be too impossible a challenge in the competitive arena. Better to focus on core competencies and value network the rest with others who are best at what they do. The challenge then lies in the integration management of scattered core skills and stages of the value chain. No easy task, but beginning to be made much more achievable by the onrush and sophistication of electronic communication.

First Virtual Corporation, a US producer of desktop multimedia, sees its world moving fast in this direction. 'We focus on only two core competencies', says founder Ralph Ungermann, 'fast and continuous innovation and building powerful partnerships . . . everything else is deliberately outsourced to "best of breed" suppliers . . . we depend on these partnerships and we'd die if we couldn't make them successful'. Hewlett Packard's Vice President, David Logan, makes a similar comment: 'one of our primary core competences is to manage alliances . . . it will become more important in the future.' In further research conducted by *The Economist*, in all 22% of executives interviewed felt that through the next decade they would increasingly find themselves working 'as part of a large network of companies'.

The corporation of the future is likely to look more like an organic enterprise rather than uniform and rigid throughout. It will leverage this new organisation form to have more flexibility. Partners can be dismissed if they fail to deliver. If the market grows, new partners can be added. If customer needs change new competences can be easily and quickly incorporated into the value-delivery system. Even supposedly 'core competencies' will be made to justify their retention inside the corporation. 'Every thing we do is up for grabs' is part of a new approach to organisation structures being adopted by some like NYNEX Cable Comms and British Airways. BA's John Patterson foresees a scenario where there is no longer any certainty about which core competencies are retained: 'if we are to stay successful in the twenty-first century we have to make sure we can organise with the best possible combination of talents and abilities'.

*Knowledge capture*

Knowledge, as Drucker has pointed out and as described in Chapter 10, becomes the driving core competence in twenty-first-century commerce, replacing land, labour and capital. Its prominence is due wholly to advances in IT which have made the capture and distillation of data into knowledge more immediate and accessible and achievable at reasonable cost and affordable investment.

Moving to a state where the right information is available with the right people at the right level of insight at the right time is now a major challenge for business generally. Only a few such as Fedex, UPS and USAA can claim to have got most of the way there already and reaping the benefits. One best-practice leader in this field is Chaparral Company, a US mini mill, that produces steel. It has achieved extraordinary levels of productivity by becoming 'a learning laboratory dedicated to knowledge creation, collection and control'. It describes its central purpose as 'tapping knowledge and translating that into enterprise-wide success'. Another emerging leader is pharmaceutical group Hoffman–LaRoche. One of its managing directors, in workshops through 1997, found there was great scope to speed up time-to-market for new products through knowledge management techniques. By mapping out what information was held where, identifying who were the key providers at the centre of informal networks, making selected other data available to key people in the chain and removing roadblocks slowing down data transmission, astonishingly up to two-thirds of the time it had taken could be saved on certain products.

Knowledge management will create value and the key organisation learning is to eliminate 'knowledge silos' where information is held by some and only selectively released on request. Logging and recording *all* the available information is an important step to realise the potential and this needs to get at both explicit data as well as the 'tacit knowledge' that's in people's heads. Powerful new software tools and 'neural networks' can help make rapid sense of masses of data, but organisation structures will likely also need to change to facilitate knowledge capture. At Unilever, for example, there is a move to reduce hierarchies and numbers of reporting levels to make it easier for ideas to flow up and down the organisation quickly. Most corporations now acknowledge that there is significant potential in improving their knowledge-capture competencies and that all parts of the organisation will need to be enlisted to make the essential and continuing contribution to the central knowledge pool.

*Disintermediation*

The potential impact of disintermediation on organisation structures and ways of doing business is immense. It displaces the intermediary, cuts out the broker, agent, distributor and retailer from the supply chain. It will force a search for a new identity, a revised role in the market place and a significant shift in the value-added the intermediary aims to bring. It will also challenge manufacturers and suppliers to build their own direct supply lines to customers. Can they effectively forward integrate in this way, can they master new competencies in customer management and home delivery? It further demands new skills and systems to make direct electronic selling and communication effective and satisfying. Does a company have the in-built flexibility and agility to learn these new approaches and respond quickly enough?

The competitive environment will give no respite and those seeing the disintermediation phenomenon coming to their market place will need to take a total view of how it affects their organisation. Honest appraisal may well confirm that for some companies the new competences required to take advantage of the situation are just a step too far to take on their own. This is where 'value networks' (see Figure 6.2) can become part of the solution path, putting together the select group of partners, allies and outsourced operations that can still enable an effective response to be made.

The growing sophistication in information management and communication technologies will continue to encourage and facilitate direct exchange between sellers and buyers. As skills and systems develop and mature so disintermediation threats and opportunities will become more widespread. Managing that effectively, as suggested in many places in this book, will become a key factor for business success in the next decade.

*Convergence*

There is much discussion about the coming together of the computer, content and communication industries (see Chapter 10), but internally, within the organisation, perhaps one of the key dynamics is the relationship and interface between two particular departments. Historically, IT and Marketing have been disconnected – one immersed in technical know-how and systems maintenance, the other with a mostly outward-only orientation towards markets and customers. That's been acceptable up to a point but communication and collaboration between the two departments has been limited. Different perspectives and even different cultural orientations have kept the two in silos. There's been little 'convergence'.

The new world of electronic commerce demands a much more integrated approach. IT is no longer just an enabler, it becomes a primary route to market. It can't stay in a back office concerning itself with systems maintenance or arcane developments. It has to come out directly into the market place. It has to meet up with Marketing. In similar fashion, marketers must become technically literate. They cannot just concentrate on being creative and developing innovative advertising campaigns. They have to come to terms with the electronic information age and learn how to be innovative in that environment too.

Such a coming together may not be easy. Marketers have typically eschewed technical literacy, IT departments have often had little time for what they perceive as something ephemeral and insubstantial and sometimes hard to see the immediate pay-back. But electronic commerce will force this organisation issue to be resolved. Joint action teams, secondments and learning from each other will be critical if an organisation is to effectively respond.

*Organisation agility*

To quote again from the 1998 *Economist* survey, 85% of companies in the research cited 'flexible organisational structure' as a key factor and challenge for the next decade. Larger organisations appear to see this as especially important, recognising their current struggles to respond quickly to new market place issues. As Professors Bartlett and Ghoshal have pointed out: 'the traditional multi-division matrix organisation such as General Motors looks out of place, a new structural paradigm is called for'.

At the heart of the twenty-first-century corporation the research and evidence is highlighting the requirement for 'agility'. There's a need for companies to develop an institutional capacity to effectively and speedily manage an environment of constant change. A new mindscape is required, involving how individual employees see their role in the organisation, the relationships they have with others and their alignment and commitment to the targets and strategies that are set. Agility has been described by advocates CSC as establishing the inner entrepreneur, the aptitude and energy and capability to respond quickly, an unleashing of the human spirit for enterprise to meet the challenges ahead.

In order to create this environment new approaches to organising the company and its workforce are appearing. There's an emphasis on focusing down on a few core business areas, divesting or demerging activities which might distract or don't naturally fit a parent's core competences. There's increasing emphasis on delegation or more fashionably empowerment, pushing out responsibilities and accountabilities and encouraging local decision-making. There's a recognition of the flexibility that small teams can provide where there are limited hierarchies and reporting levels, easy communication and a clearer shared mission and purpose.

Sheer scale and size of organisation appear increasingly irrelevant in this new age and if anything a handicap rather than a disadvantage. 'There is a growing sense', as pointed out in a recent *Financial Times* article, 'that big companies do not have what it takes to compete in the twenty first century, they struggle to develop the flexibility, rapid response, and openness to innovation that's required, they no longer attract the best people'. And as Professor Michael Porter, once a proponent of scale as a source of advantage, has recently acknowledged: 'the big company of today is not being defeated by another big company, but by small companies. As information flows very rapidly, as companies compete globally, scale is no longer a competitive advantage.'

Companies, therefore, that are rushing into multi-billion-dollar mega mergers should take heed. Economies of scale may be illusory and there may be a hidden cost to that scale that can seriously handicap the corporation's ability to respond to new market challenges. Driving to a state of 'organisation agility' may be the only means to deal with rapid change on an effective and sustainable basis.

**Strategic Response**

In order to deal with these various organisation challenges, the very first step is to develop a long-term game plan and strategy for electronic commerce. Given the various options available, ranging from embracing the new electronic age through to bucking the trend and defending vigorously the status quo, which route forward is the company to take? What will maximise shareholder value? What future targets should the company set in its chosen markets with its major customers? What dream or vision can most propel the company forward? What stretching reality is achievable and what's it going to take to get there?

Earlier chapters described two key tests for the electronic age – the ES Test and the Manufacturer's Test. These give guidance as to the options, they help understand how specific market places are evolving and how many customers will be looking for what degree of change. They also help establish how capable the company is to deal with the upcoming challenges and whether it has the strength to resist and pioneer or must go with the flow.

But these test inputs need to be brought together. There needs to be a developed and cohesive understanding of what it all means. Stakes need to be put in the ground, horizons extended, targets set, funds and resources allocated and the commitment instilled. Competitors must be assessed, customers' reasons for buying from one rather than another evaluated, opportunities for differentiation reviewed and sources of competitive advantage determined (Figure 15.1).

There can be no avoiding the need to set a clear strategy, conduct the market analysis and appraisal and get to grips with the electronic future. As a number of studies have pointed out, only in this way can a company hope to achieve the growth it's looking for. In a 1997 *Sunday Times/Coopers & Lybrand* review of what factors enabled companies to keep growing and sustaining their success the key that emerged was 'a strong focus on clear strategy-making'. In another study by the British Quality of Management association, the attribute that most distinguished Britain's best companies like British Airways and Marks & Spencer was their strategic thinking and the way the leadership of the company harnessed that to drive their business. And in the most recent *Fortune* magazine assessment of 'Most Admired US corporations', developing the

**Figure 15.1** Key Elements of Effective Strategy-Making

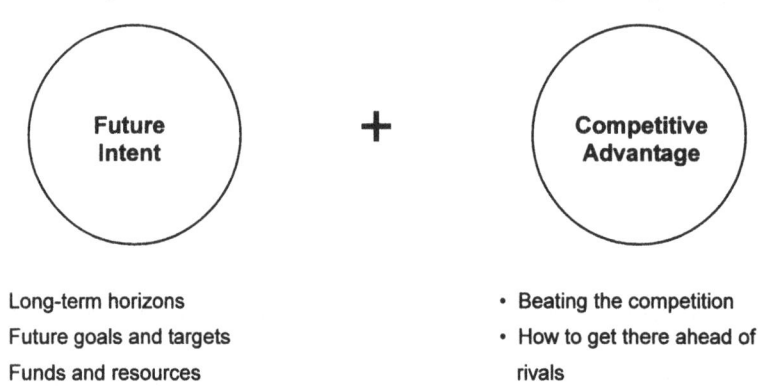

- Long-term horizons
- Future goals and targets
- Funds and resources

- Beating the competition
- How to get there ahead of rivals

vision to take constructive risks in the market place emerged as a hallmark setting the most admired apart from their peers and rivals.

Despite the compelling evidence, those companies that do develop effective, driving strategies are surprisingly a minority. Most in practice pay lip service to strategy, putting out plans which are often no more than extrapolations of twelve months' budgets. Few truly invest the necessary time and resource to rigorously review where the company is headed, what the opportunities are five, ten years out, what knowledge gaps there are in understanding what core competitors' future strategies and intentions are, what customers are really going to want in the future, which product lines are making money and adding shareholder value and which unlikely ever to contribute.

The research described in my 1997 book *Strategy in Crisis* shows that many organisations are caught in a vicious circle (Figure 15.2) where strategy-making is a lost art. It's often low on agendas, displaced by short-term financial performance goals, or pressing operational issues like Y2K (year 2000 systems compatibility). It's often replaced by an internal orientation toward process change or restructuring. The basic strategy skills get submerged, corporate planning becomes de-emphasised and there's no voice at senior executive committee or Board level counter-balancing these alternative agenda items and urging the longer-term perspective. Without the appropriate strategic framework there's the risk of knee-jerk reaction to events or market changes, actions taken in a vacuum and the organisation appearing rudder-less, vulnerable to the storm-tossed seas instead of sailing a purposeful course.

*Setting the Strategy, Mobilising the Organisation* 237

**Figure 15.2** Strategy-Making – Caught in a Vicious Circle

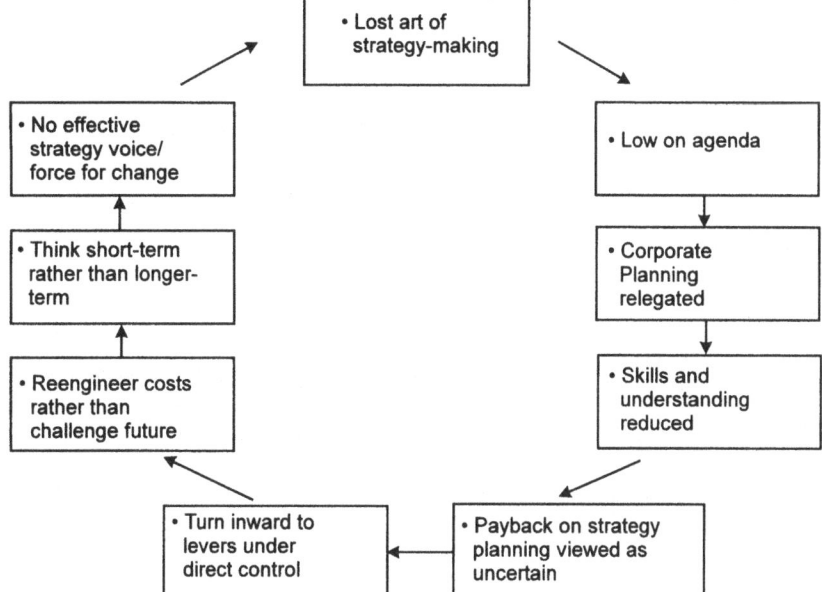

Companies that are visionary, that do embrace strategy – *they* set out to make sense of their environment. They've discovered strategy's power and potential. They've arrived at a balance between all the short-term agenda items and the longer-term health of the corporation. They know they can't do without it. As the Chairman of UPS has pointed out:

'nowadays we are making ever bigger bets in our investments in new technology so we can't afford to back one direction and then find out five years later it was wrong . . . putting time and effort into strategy-making is vital . . . strategy is our most important management issue and opportunity.'

Putting strategy first and developing it rigorously unquestionably helps a company understand its markets thoroughly, evaluate its options and determine the most appropriate and leveraged course of action. In extensive research by Collins and Porras (Figure 15.3), companies that did this out-performed both their rivals and the general stock market many times over. They demonstrated that this is a key test for would-be shareholders. If the company is to add sustainable value it must show it has the capability and skills to think and plan strategically and set its long-term game plan.

**Figure 15.3** It is Companies That Do Plan their Way Through Which Emerge as Winners

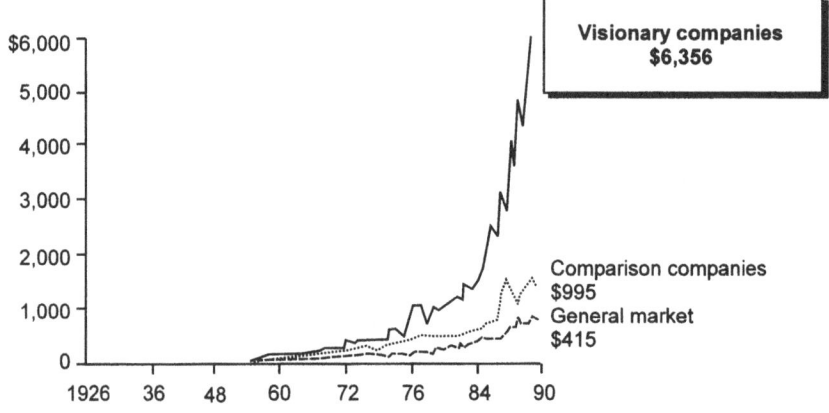

*Source*: Collins and Porras.

### People-empowered implementation

As companies begin to get their minds around their e-strategy, so they will get a much better perspective on what it's going to take to achieve their new ambitions. The organisation changes required will inevitably be considerable as the impacts caused by electronic commerce are so widespread. Each of the issues described at the beginning of this chapter, from Digitisation and Virtualisation through Disintermediation and Convergence will create their own distinct challenges and opportunities. Each needs to be separately thought through and the appropriate structural, system and skill response determined. It's a daunting task. So much change, encompassing so much of what the organisation does. Few parts of the operation can be left untouched. The amount of time, effort and application to see things through and make the changes will likely appropriate significant resource, funds and energy.

Strategy-making can help with navigating the path through this. It can establish priorities, mark out what's important, highlight the key success factors (e.g. the main source(s) of competitive advantage) and keep things focused. But the people in the organisation are still going to have a lot to deal with, they are still going to be expected to go the extra mile, take on new challenges and treat them as part of the everyday job.

The key is to engender a collective effort, mobilising employees, suppliers, value partners and others behind the new strategy so that the whole is capable of achieving significantly more than its individual parts. There are ways in which senior management teams can engage the organisation so that the workforce are excited about the future, rather than daunted by it, energised to go the extra mile rather than enervated by the prospect. A supportive working environment can be set up which encourages open communication, sharing of ideas and problems, promotes co-operation, rewards hard work and harnesses the total energy and effort. We can describe the holistic approach required as the Sharing Plan (Figure 15.4).

The goal here is to develop that powerful sense of ownership and personal responsibility among the workforce. The opportunity is to create that passion and excitement that is often seen around successful founding entrepreneurs or around the more driven and inspired organisations such as Wal-Mart, The Limited, Nike and Microsoft. It's about generating a genuine enthusiasm for the job that finds Home Depot associates saying: 'by giving 100 per cent to my customers, I'm helping to write the success story of this rapidly expanding industry'.

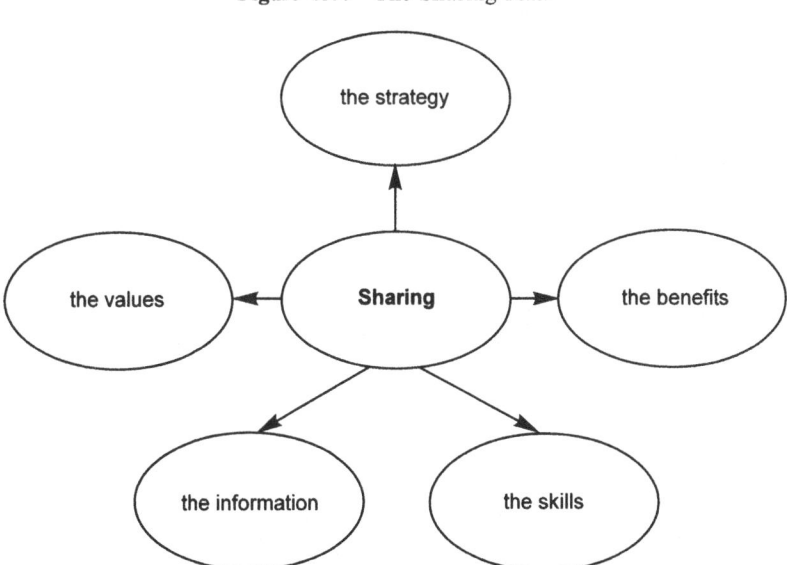

**Figure 15.4** The Sharing Plan

*Sharing the strategy*

Charles Handy in his book *Age of Unreason* explored the attributes of successful companies. His analysis, corroborated by others, is that any strategy that's set must be 'involving'. It must be developed into a vision 'that gives point to the work of others'. Business leaders have a responsibility to communicate something to their employees that is a compelling and believable vision of where they want the company to go and what role and contributions the workforce has in getting the company there.

As Anita Roddick of Bodyshop puts it so succinctly:

'You have to find ways to grab people's imagination. Most businesses focus all the time on profits, profits, profits. I think that is deeply boring. You want people to feel they are doing something important.'

There's a requirement to translate the strategy into something that even the most junior employee finds exciting and engaging. So Toyota's formal strategy may be based on product excellence but in its communication to the workforce it's put in the context of 'Beat Mercedes'. Avis may have determined its strategy around service leadership, but it was communicated in a way that rallied the workforce: 'we may be No. 2 but we try harder'. McDonald's may focus on superb service but it landed on the expression QVSC (Quality Value Service Cleanliness) as a simple mnemonic that its employees could remember, be measured against and deliver on.

The strategy provides the platform to set the direction. If the strategy-setting is rigorous enough it should provide plenty of opportunity to be translated into motivating goals and in a language and context employees can readily buy into and accept. It must 'touch the spot', be informal and honest, come from the heart as well as the head. 'If people are to put out the extraordinary effort we now require to realise corporate targets then they must be able to identify with them and share in the ideas and goals they represent' (Jack Welch of GE).

*Sharing the values*

If it's the employees who can make the difference, then why treat them as second-class citizens? Why have separate management canteens (a peculiarly European phenomenon), reserved car-parking spaces, separate floors in the building, restrictions on information flow? These things are symbols of divide-and-rule and they cut people off from listening, interacting, sharing and communicating with the rest of the company. They are just the kind of needless barriers that get in the way of effectively sharing the values and the strategy that can unite the organisation.

How much better to have more open-plan offices or a 'no-closed-doors' policy, as at Nike. At Solectron, a contract manufacturer that has won the Malcolm Baldridge Award, neither the chairman nor the CEO has a private office. At Asda, the UK supermarket chain, when Archie Norman was CEO, he abolished chauffeur-driven-cars – except one. He kept it, not for the senior officers of the company, but for 'the employee of the week'. Each week the employee, whoever they were, who came up with the best idea for the company to improve things, was awarded the car for the week, with the chauffeur, to take them about wherever they needed to go.

This was a simple but highly symbolic act which quickly entered company mythology and helped change the way people felt about their employers. They appreciated the sharing, the 'we're all in this together', the removal of barriers. Not surprisingly it helped, with other things, to create a tremendous sense of shared purpose and mission.

*Sharing the information*

Why not openly share detailed financial and other information throughout the company? If the feeling is that employees cannot be trusted with that information then that is surely a sign that there is no common purpose and effective shared values and beliefs. Jack Welch at GE has a policy of open communication and explanation as to how the company is doing, what it's doing well and what it's doing badly. At Lincoln Electric, 'information is shared with all employees regarding the financial and market position of the company'. At Advanced Micro Devices every employee has been specifically trained to use the company's IT network to access information about all aspects of the company's performance.

Open information-sharing makes employees naturally feel more involved and more responsible for the company's performance. How much better to hear it from your department manager and be able to discuss and think through what its implications are than read about it with the rest of the world when the third-quarter results are published in the press. An openness and readiness to share and trust employees with the information has to be a major reinforcement to the strategy and values platform just described.

*Sharing the skills*

People cannot be expected to make the difference unless of course they have all the skills required to do the job. And as the competitive environment becomes more complex and demanding, so there is a growing imperative to continue to improve the skill base and give employees from all parts of the organisation a chance to continue to learn and gain in experience and capability.

There is a constant battle in most companies today between, typically, the human resources function who are the usual sponsors of better training, vs the company's financial engineers who are concerned with reducing costs wherever an activity or investment is discretionary. It's part of the ongoing short-term vs. long-term debate which continues to pressurise companies into sacrificing the long-term for the sake of the short-term earnings.

However, what research has shown – conclusively – is that companies that do treat training as a priority and invest heavily in it reap benefits and dividends often far more quickly than they expect. Rather than complain about not having enough good people and not having the right skills in the company to meet the new global market challenge, advocates of increased training would argue that people do have the innate core competencies and if only they are given requisite training it will unleash a new set of energies and abilities.

Some companies have grasped this particular nettle with vigour and investment. McDonald's, for example, has established its 'hamburger university' which rigorously trains all its employees and recognises their achievements in QVSC. Toyota will deliberately take people out of the manufacturing plant and put them in the salesforce for a period to experience first-hand what it takes to achieve total customer satisfaction, understanding how important it is for them to get things right. And there is the Raychem college of further education which was the particular initiative of founder Paul Cook and which offers a substantial programme of courses open to all employees, not just related to business and management but to help improve more generally 'people's capacity for personal growth'.

These companies and others like Procter & Gamble, Gillette, Publix Supermarkets and Eddie Bauer, featured in the book *The 100 Best Companies to Work for in America*, have committed to their workforce in this way. They have seen the benefits that can come through from a more highly skilled and thereby more motivated workforce. It does take investment but it is certainly one of the keys in helping and getting people ready to put out the extraordinary commitment that's now required.

*Sharing the benefits*

There are a number of best-practice examples:

- The John Lewis Partnership is the largest employee-owned firm in the world. *All* profits are distributed as a bonus to all employees. Sales have grown consistently and have achieved a compound rate of over 12 per cent. It has performed better than all its major rivals. The partnership continues to perform outstandingly well today.
- Among the top-performing stocks in the USA many like Southwest Airlines, Wal-Mart, Intel, Microsoft, Nordstrom and Compaq appear on the 'Employee Ownership List' (a listing of 1000 companies where employees own > 4% of the stock).
- 91 per cent of listed Japanese companies have employee stock ownership plans in which > 50% of the workforce are eligible and participate.
- Levi Strauss announced in 1996 that it would pay each of its 37,000 employees one year's salary as a bonus to hail the new millennium. News of the millennium bonus sparked wild celebrations at Levi Strauss, according to press reports, and predictable gloom at rivals!

    'It's about sharing our success with our employees . . . we all want a company that our people are proud of and committed to, where all employees have an opportunity to contribute, learn, grow and advance . . . we want our people to feel respected, treated fairly, listened to and involved. Above all we want satisfaction from accomplishments and friendships, balanced personal and professional lives and have fun in our endeavours.'

\* \* \*

This chapter has been all about coming to terms with the new era of electronic commerce. It indicates some possible paths and solutions that may help companies move forward. It sets out a framework for strategy-making that discourages anything less than a thorough and rigorous approach. It provides a tool kit for getting the workforce involved and mobilised behind a future game plan and ready to go the extra mile to deliver it.

The organisation impacts are going to be wide-ranging and only a dedicated and committed approach can ensure they are effectively dealt with. Of a necessity new structures will be required, new customer management skills will need to be learnt, new knowledge capture systems developed. The whole will take time, resource and considerable application. But success in any field never comes easily. Getting the workforce to make the difference means battling with the wills and egos of individuals, managing their natural concerns and uncertainties and some wavering of confidence as to whether a team truly has the capability and skills for the battles ahead.

What's critical is to acknowledge the magnitude of the change management challenge. It is difficult, it does require a specific task force to handle it. It will need to be led consistently and proactively by the senior people in the company. What's also important is to acknowledge the success stories among those companies who are succeeding. There's great advantage in learning from what they have done and recognising the effort they take in sharing and involving and communicating. Harnessing the energies of the whole organisation will be pivotal. It can provide the platform to inspire a common purpose that drives relentlessly toward lasting e-success.

# Appendix 1:
# The Retailer Dilemma – Can Shops Still be Profitable?

Retailers generally have been under increasing pressure through the 1990s. In many sectors, margins and sales densities seem to have peaked and retailers have been forced to look elsewhere for continued growth. Now electronic commerce is rushing toward them the need for continued change is even more important. Yet just when they need to be at their most flexible and adaptable they find themselves hamstrung and hesitant. Their commitment to their physical space is so ingrained, it's so woven into the fabric of what they are and what they do that they're finding it extraordinarily difficult to stand back and consider all their strategic options openly and evenly. Their temptation is to make the real estate continue to work for them. But as we've seen that may be achievable on some products and services with some consumers, but it won't work in many other situations.

This retailer dilemma is considerable. Invest in space or embrace the new electronic age? It's a clash between the need to change and the vested interest and commitment to physical real estate. It's making many hesitant and encouraging others to still bury their heads in the sand and hope that electronic commerce will never develop and mature. Such a policy is doomed and it's clear retailers must confront this 'clash of interests' and learn how to move through it. They've long had a history of adapting to survive and grow and the next few years will present one of their biggest challenges yet. Just how significant is this real estate commitment? To what extent are retailers tied to the land? What other pressures are they facing? Will they be forced to experiment and innovate or can they comfortably carry on as they are?

In some ways this retailer dilemma is similar to the predicament in which IBM found itself in the early 1980s. IBM was committed to mainframes and saw this as the future of the computer age but it was soon confronted by new technologies that were making small-unit PCs increasingly powerful. Stick with mainframes or embrace the PC? Try to do both? Mainframes at the time so dominated IBM culture and future planning, it was so fundamental a part of what they did that the organisation just found it very hard to think objectively about PCs and react effectively to the early signs in the market of a shift in customer interest toward them.

Even though IBM records show the company was probably the inventors of the PC they of course stuck with what they thought they knew best. Their marketing of the PC was half-hearted and peripheral. They hoped it would go away. In the meantime other unencumbered players like Apple came along and cleaned up.

Ten to fifteen years later and the successful PC manufacturers are now facing their own distribution dilemma. While Compaq continues to sell principally through retailers and its own salesforce, Dell is growing at an enormous rate taking market share by dealing direct with the purchaser. Dell bypasses the traditional channels and is aggressively taking advantage of the electronic media. Should Compaq, Hewlett Packard and others follow? Do they too abandon traditional ways of operating and rush into the new way of doing business? Do they have the objectivity to see the nascent shifts in consumer habits of buying what is increasingly becoming a commodity? Or are they so ingrained in 'their way of doing business' that they'll continue as they are while Dell and a pack of smaller rivals come in and clean up?

Will retailers take these history lessons on board? Will they recognise they are about to face the same major channel and technology threats? Will they confront the shifts in consumer shopping interests and habits? Will they evolve or will they, too, be finally caught out by new entrants with no historic baggage and infrastructure? Certainly retailers show no signs of slowing down their commitment to their physical real estate and they continue to behave as if that traditional way of doing business will endure forever.

Typical retailer reactions to these challenges in early 1998 could be summarised by this selection of quotes:

- 'Well, yes, it might happen but right now it is off our planning horizons.'
- 'Shops have been around for 2000 years, we're not going to be replaced.' [Dinosaurs probably said the same thing!]
- 'We're not worried it's just an additional channel and it will all be incremental business anyway.'

That retailers are locked into their real estate is evident from the amount of money they have invested and continue to invest in land and buildings.

## The Retailer Dilemma

Balance sheets of US retailers in 1997 showed some $250bn of fixed investment in land and buildings. The UK equivalent figure was nearly $50bn. And that excludes banking, other retail financial services, restaurants and fast-food outlets.

The amount of physical space retailers are committed to also continues to grow. For example, between 1987 and 1996 the number of supermarket outlets operated by supermarket chain Sainsbury in the UK grew from 54 to 227. This involved more than trebling the sales space from c.7 mn to more than 25 mn sq. feet. Taking a US example, the number of domestic stores owned by Wal-Mart has grown from 980 to nearly 2000 over the same period with sales space again more than trebling this time from 58 mn sq. feet to more than 180 mn sq. feet.

There is little sign that this rate of investment is likely to slow down, with all major retail chains continuing to make announcements of their plans to grow their space and to do so fast:

- In its 1997 annual report Wal-Mart listed total planned capital expenditures for 1998 alone of $3bn, targeting 50 new Wal-Mart stores, 100 supercentres, 5 to 10 new SAM's clubs – and all that just in the USA aside from continued space expansion overseas.
- In November 1997, Marks & Spencer announced a major new space expansion involving $3.5bn over three years opening more space both domestically in the UK and overseas adding some two mn more sq. feet to take its total retail space to 10.5 mn sq. feet.
- Thorntons, the UK chocolate manufacturers and retailers, proudly revealed the following space expansion programme through till 2001 more than doubling square footage from 1997 levels (Table A.1).
- 

Table A1.1  Four-Year Stores Plan for Thorntons, 1998–2001

|  | 1998 | 1999 | 2000 | 2001 |
|---|---|---|---|---|
| New openings | 52 | 57 | 56 | 42 |
| New sq. footage | 31,000 | 33,000 | 30,000 | 21,000 |

Home Depot opened its 500th store in 1996 and has boldly announced plans to reach 1300 by year 2001. It has grown square footage at a rate exceeding 20% each year through the mid-1990s (Figure A1.1) and plans to continue at that rate, even though in 1997 actual sales per sq. foot grew at only 3%. The company says there is still plenty of room to grow.

## Appendix 1

**Figure A1.1** Home Depot Space Expansion, 1991–6

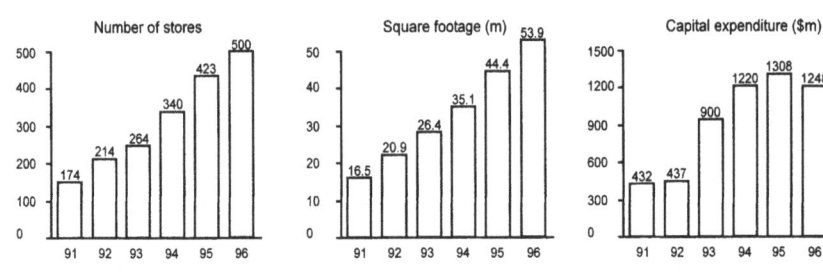

*Source*: *Annual Report*.

- In the UK alone, some six major new out-of-town shopping centres were scheduled to open between 1997 and the end of the century providing an additional 5 mn sq. feet of retail selling space (Table A1.2).

**Table A1.2** UK Major New Shopping Centres, 1997–9

| Centre | Opening | Sales area ('000 sq ft) | Anchor tenants |
|---|---|---|---|
| White Rose | March 1997 | 650 | Sainsbury, Debenhams |
| Cribbs Causeway | March 1998 | 725 | John Lewis, Marks & Spencer |
| Trafford Park | Winter 1998 | 1000 | Selfridges, Debenhams |
| Braehead Park | Winter 1998 | 696 | Marks & Spencer, Sainsbury |
| Bluewater Park | April 1999 | 1454 | John Lewis, Marks & Spencer |
| White City | Winter 1999 | 620 | Not finalised |

*Source*: Corporate Intelligence on Retailing.

A trawl through the *Financial Times* almost any month of the year provides a stark illustration of the pace at which retail chains are continuing to add space. For example, in June 1997 Boots were announcing plans to open 30 new stores in the Netherlands with each store three–four times larger than its existing average store. Carrefour, the French retailer, was talking up its plans to double the number of its stores to 500 world-wide over the following five years. Eroski in Spain said it planned to open 30 new hypermarkets and 80 new supermarkets by the year 2000. Liedl & Schwartz, the German discount grocery chain, committed to open 30 more units in Germany and Scandinavia over the following twelve months. And these were just the more significant of the store space announcements that month!

Retailers seem incapable just now of jumping off this new space bandwagon. They seem reluctant to pause to take stock of electronic commerce, how it will impact and whether they should be modifying their plans. What makes their current headlong rush even more challengeable is the long-term commitments they have to make to the space they buy. To get the pay back on the massive investments in land and buildings retailers often have to set 20–25-year time horizons. Leases are invariably taken for that duration – sometimes with upward-only rent reviews! – and depreciation schedules usually assume life expectancies from 10 to up to 45 years! The size of the investments in land, buildings, fixtures and fittings can easily top $20 million and is sometimes considerably more than that involving further investment in infrastructure, road networks to ease access and compensations to the local community for disruption and disturbance.

The size of these commitments has always represented a significant gamble but through the 1980s, especially as consumer spending and confidence soared, they seemed a relatively safe move. There did seem to be plenty of demand for these new large stores, many areas had yet to be penetrated, there seemed to be opportunities for further improving margins and sales per sq. foot. It was as close as business is ever going to get to a one-way move. But since the early 1990s recession, the competitive environment has toughened considerably and now with electronic commerce coming along this historic way for retailers to build their business no longer appears as secure or as financially attractive.

Going forward the situation for retailers is exacerbated by the growing pressures on their business. These make facing up to the electronic shopping resolution even more challenging but equally all the more important to see if it can provide new sources of real growth and advantage. Current pressures include:

- Margins are already wafer thin
- Sales densities appear to have plateaued
- There is saturation now in the number of shops
- The major supply-chain efficiencies have largely been captured
- For supermarkets specifically, there is a growing shift from eating in to eating out.

## Margins are wafer thin

US and European retailers get by now on net profit margins of just 2–3%. Wal-Mart's net margins, for example, in 1993 were just 3% and with its recent acquisition of the Wertkauf supermarket chain in Germany it is entering a market where average profit margins are only 0.7%, dragged down in recent years by aggressive price discounting by the likes of Aldi. Only in the UK do retailers continue to post higher margins of around 4 or 5% but these, as we shall see, look to have peaked.

As Bill Gates recently put it when addressing a retail conference in the US: 'I'm in a 20 to 30% margin business, you're in a 1% margin business. Not too many of us are trying to figure out how to get into 1% margin businesses.'

## Sales densities appear to have plateaued

US retail sales per sq. foot between 1983 and 1995 grew in real terms at only 0.4% and since the boom of 1987 have actually stayed flat to marginally declining (Figure A1.2).

The UK has seen a very similar pattern with real sales per square foot declining by 3.9% between 1990 and 1996. While absolute sales did continue to grow overall across this period, it was offset by significantly faster growth in the amount of selling space leading to inevitable cannibalisation and dilution of sales per sq. foot effectiveness. This sales decline is compounded by pricing pressures such that prices are static to falling in retail sectors like food, despite a 2–3% rise in overall inflation each year.

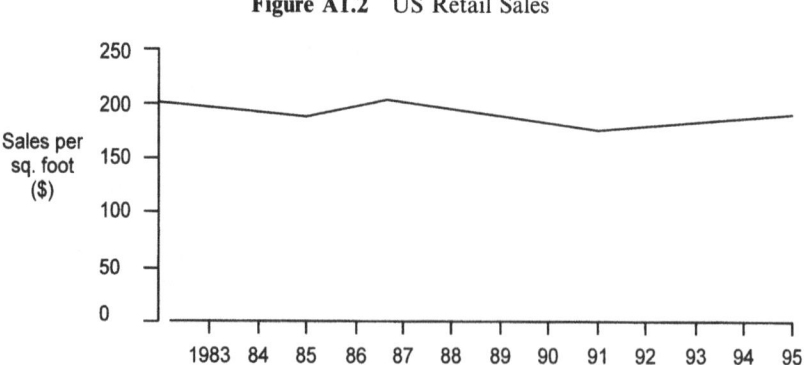

**Figure A1.2** US Retail Sales

## Saturation in the number of shops

In the home improvement sector in the USA, analysts estimate that the country can successfully support about 2000 'big-box' stores. Yet at current rates of growth, Home Depot and Lowe's – the two leaders in the field – will by themselves have that number of stores by the year 2001. By that time every market with a population of 50,000 people or more will have *at least* two big-box stores each! In similar fashion, Marks & Spencer in the UK is continuing to add retail space but by 1997 it already had saturated most of the country. Every urban centre or catchment area with a population of more than 100,000 had its own M&S store. There were only six relatively small areas without. Yet as M&S continues to add space can it expect it all to be incremental, isn't there likely to be some dilution creeping in?

Most retail commentators acknowledge much of western Europe and the USA is already potentially saturated with shops and that can only lead to reductions in effective sales per sq. foot and continued pressure on margins:

- 'As the big superstores near saturation of the major urban areas they are being forced to move in on smaller communities, pushing out the "mom and pops" but also finding it harder to maintain sales and margin levels.'
  
  (*Oxford University Institute of Retail Management*)

- 'There is too much floorspace sharing too little spending. But retailers are not shedding space but opening yet more of it.'
  
  (*Verdict retail report*)

- 'We are overshopped. Do we need a world-beating food industry or a world-beating spending industry? We don't need more supermarkets.'
  
  (*Archie Norman, Chairman of Asda*)

## The major supply-chain efficiencies have largely been captured

When the early 1990s' recession began, retailers redoubled their efforts to improve the efficiency of their operations. They worked across the whole supply chain involving suppliers in cost-reduction programmes and partnerships to improve transportation, warehousing and logistics. They also looked at their store operations and improved staff scheduling and productivity. They invested heavily in total in new systems like EPOS that gave them better information to manage stock levels and availability and generate the maximum utilisation and leverage of shelf space.

The work was intense during the first half of the 1990s and the innovation was considerable. But most of the major chains have now put much of these supply-chain efficiencies in place. They've put in EPOS, established EDI links with suppliers, outsourced where they can, fine-tuned the way the stores operate, developed private label and that too has reached saturation levels in many chains (in Sainsbury it accounts for c. 50% of all sales). Much of the cost reduction benefits have now been achieved.

At the gross margin level things look to have plateaued at around the 30% mark. Sales per employee, like sales per sq. foot, have flattened out. While there are no doubt further initiatives in the pipeline in many retail organisations, it looks like the big wins have already been taken and what's left is a series of smaller incremental adjustments which by definition will have less impact on profitability.

**For supermarkets specifically, there's been a change in eating habits**

The pressures just described are affecting all retail groups but food supermarkets are facing an additional threat. That is the shift in consumer spending habits. As general levels of affluence or disposable income have risen and as 'time poverty' makes life-styles more and more difficult so consumers are looking to spend on eating out rather than eating in. They are switching their spending away from the supermarket to restaurants and other catering outlets (Table A1.3).

As *Business Week* highlighted in its 1998 survey of the prospects for different industry sectors: 'for the past 25 years the traditional

Table A1.3  Trends in Consumer Spending, 1975–95

|  | (% of total) | | | | |
| --- | --- | --- | --- | --- | --- |
|  | 1975 | 1980 | 1985 | 1990 | 1995 |
| Food | 18.8 | 17.1 | 14.1 | 12.0 | 10.7 |
| Drink and tobacco | 11.6 | 10.7 | 10.4 | 8.6 | 8.3 |
| Catering | 5.3 | 5.8 | 5.7 | 8.6 | 8.7 |
| Entertainment | 9.1 | 9.2 | 9.0 | 9.7 | 10.3 |

*Source*: UK National Accounts.

supermarket has been sliding toward oblivion. Once responsible for almost two-thirds of food sales its share in recent years has fallen to only just above 50%.'

* * *

Robert Tillman, the head of Lowe's, provides us with a succinct summary of the challenges ahead for the retail industry:

> 'the competitive landscape in our industry is changing, it is inevitable and unavoidable. For many it will probably mean failure. I believe retail saturation combined with growth coming in electronic commerce represents the biggest challenge facing our industry today. How do we grow shareholder value when we can no longer grow what we are doing and what is growing is something we are unfamiliar with?'

Tillman has argued strongly that there is an urgent need for retailers to really get close to their changing consumer needs and adapt their proposition accordingly. 'Stay current with the wave of technological change,' he argues 'or get swept away by it . . . planning and positioning for tomorrow must relentlessly assess and challenge today's status quo.'

Retailers must come to grips with the need for the change and not let their vested interests in space and physical real estate get in the way. They must try to shake off the historic baggage and look objectively at their future and what it will take to stay in tune with their customers. If they don't they could lose out big time like **IBM** did. If they do get themselves free from their physical shackles then they can start to play an active and leading part in setting the shape of consumer shopping in the future.

# Appendix 2: Retail Banking Case Study

Bill Gates has described retail bank branches as 'dinosaurs of the digital age'. A 1997 report by the London Centre for Financial Innovation commented that 'the growth in electronic technology could lead to many of today's cherished banking institutions being replaced . . . in their stead may be a new generation of electronic service providers and screen-based markets'. Citicorp chairman John Reed has warned that in the next decade retail banks may become nothing more than 'lines of code in a big computer network'.

As the electronic infrastructure builds, the role of the bank branch is under increasing threat. Consumers and providers can deal directly with each other. Telephones, faxes, e-mail and the Internet all make it easy to carry out transactions from home or office. What possible value-added can the branch continue to provide? As we rush toward the next century we're beginning to see the traditional distribution system for retail financial services becoming obsolete. It's suffering from 'creeping disintermediation'.

Of course, this is not just a retail banking phenomenon. Financial services generally seem vulnerable. One reason is the intangible, 'virtual' nature of a financial transaction. Taking the ES Test, product characteristics have a high 'intellect' score – it's a rational rather than emotional or physically involving purchase. Consumer familiarity and comfort with basic transactions like withdrawing and depositing cash is high and there are typically well-developed brand values offering reassurance and reliability. Combine this with the latest consumer demand for convenience and a readiness to experiment and retail banking begins to score very high on 'ES potential'.

In fact disintermediation in banking is not some new trend in response to the Internet. It started back in the early 1980s when technology first presented new opportunities in customer service. There was a boom in the use of Automatic Teller Machines (ATMs) and the plastic card that went with them. People enjoyed having their personal password or pin code and the flexibility to withdraw cash when they wanted to, especially since at the time bank branches often opened only Monday–Friday 10 till 3.

Alongside the widespread use of ATMs, there have been other developments in electronic technology which have further brought into question the future role of the branch. Telephone banking has grown to substantial levels, especially in the USA where some 40% of households now conduct some or all of their banking in that way. PC banking is now being increasingly made available with software from Quicken's Intuit for example allowing users to log on with nominated financial services providers to pay bills, move money between accounts, buy basic insurance and reconcile balance and cash statements. Payments generally are now available through both credit as well as debit cards eliminating the need for paper cheques and expensive cheque clearing systems with transfers now able to be carried out automatically.

As consumers become comfortable banking electronically so the disintermediation threat will grow. Indeed Gavin Shreeve, the Director General of the Chartered Institute of Bankers, commented in a recent speech that: 'there is no longer a need for bricks and mortar to sell and deliver financial services.' Retail banks are faced therefore with either becoming 'the next dinosaurs' or responding proactively to the new challenges. The response doesn't necessarily involve abandoning all the real estate and rushing onto the Internet. But it does involve among other things taking the ES Test and identifying what a bank's own target consumers are actually looking for. To what extent do they still want face-to-face interactions, are they content to do *all* banking electronically, how many are determinedly after the convenience of telephone/PC banking?

Certainly one response is to reappraise the number of branches. If target consumer demand *is* shifting electronically then adopting a transition policy of closing branches as and when appropriate would be advisable. Not only does this make the necessary strategic shift but there are obvious cost saving benefits as well. To date, however, there's been only limited action by banks in this direction (Figure A2.1). Sometimes there are national or political considerations which get in the way of a bank's restructuring, on other occasions banks fear public opinion which can create bad publicity when a local town branch closes threatening the commercial life of the community. Another excuse for inaction is fearing the loss of market share by being the first to cut and seeing the resulting business shift to other banks who've kept their branches open. But market share need not be lost if the closures were part of a deliberate strategy to retain and transition customers from physical to virtual banking with the investment, promotion and incentives required to make that transition as comfortable and desirable as possible.

**Figure A2.1** Growth and Decline in Branch Networks

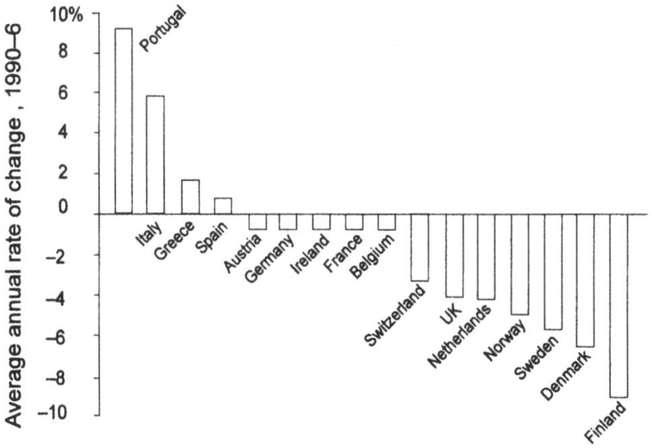

*Source*: Datamonitor European Banking Database.

In the meantime consumers aren't waiting for banks to make their minds up and sort out their market strategies. ABC1 consumers especially have been moving away from traditional branch banking (Figure A2.2). They are the consumer category typically with the most time-pressured life-styles but also with the greater PC literacy and comfort in dealing remotely. They have shown the least bank loyalty, appearing ready to switch their business to other providers who can deliver the more convenient electronic services.

For Banks who do transition from a branch to an electronic connection, the cost savings look considerable. A study by Booz Allen found that the expense ratio of costs to income for an Internet bank is 15–20%. This compares with traditional branch banking figures of 50 to 60% expense ratios. Figure A2.3 shows how this translates into substantially lower unit transaction costs for an electronic transaction. Put another way, the average cost for a bank of maintaining a customer's account in a branch network is nearly three times what a 'virtual' banking account would cost (Figure A2.4).

If banks do decide to grasp the 'electronic nettle', there are a number of options:

1. *PC private dial-up services* – proprietary software is distributed to customers which is then installed on the home or office PC. Access to the bank is effected by a modem link. As the link is a private

**Figure A2.2** US Retail Banks' 'Wallet Share': Consumers with Income > $75 k p.a., 1990–6

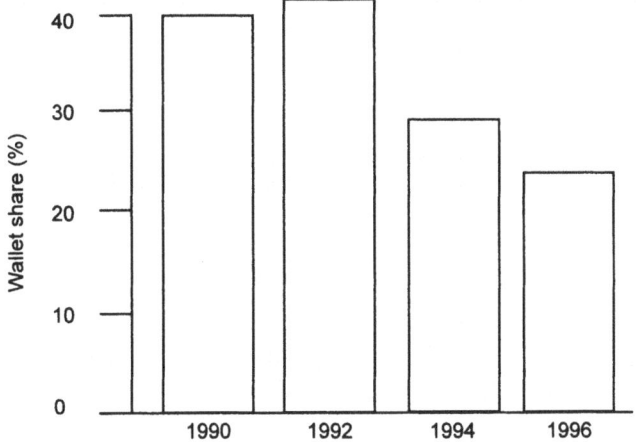

*Source*: *Computer Weekly*, Centre for Financial Innovation.

**Figure A2.3** Retail Banking Unit Transaction Costs

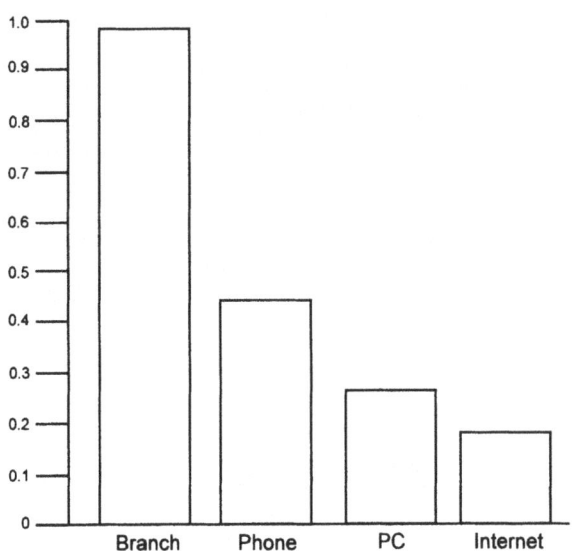

*Source*: *Business Week*, Forrester Research.

**Figure A2.4** Retail Banking: Cost of Maintaining a Current Account

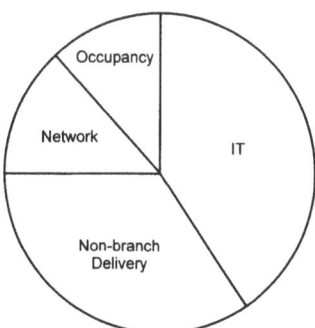

*Source*: US Banking Administration.

dedicated line it offers the customer a greater sense of security and privacy. It is typically a speedier connection than access through some third party providing the interface with the on-line service. This route has been the preferred option of most retail banks to date, so in the UK, NatWest, Midland, Barclays, Citibank, Bank of Scotland and Nationwide have all trialled this approach.

2. *Managed networks* – here a bank makes use of a network operated by another party. An example is Lloyds TSB's on-line bank which makes use of the CompuServe Internet service network. Another example comes from the USA where a consortium of fifteen banks have linked up with IBM to provide their services on-line via the IBM global network.

3. *Internet* – a bank establishes its own dedicated web site and invites customers to access it and conduct transactions through it. Only a few banks such as Royal Bank of Scotland and Barclays in the UK have trialled this approach as consumers are still cautious about security on the net (though as we saw in Chapter 11 these concerns are largely misplaced).

4. *TV-based services* – the arrival of digital TV services will probably see a rapid increase in the provision of banking services to the home or small office. Bank-account information can be delivered direct onto the TV screen via a dedicated bank TV broadcast. Interaction with the bank can be via the telephone line. Again, issues of privacy and security will need to be managed to reassure prospective customers.

5. *Interactive kiosks* – as retail supermarkets look to reinforce their own consumer-value proposition by extending the 'one-stop' range of services, so they are increasingly interested in providing basic banking services. These could be provided through a kiosk or private booth utilising touch-screen technology and enabling customers to carry out transactions and interface with a bank call centre for advice.

These options are not mutually exclusive. Banks can viably establish an electronic strategy which embraces some or all of these routes to market. Indeed a rigorous customer segmentation analysis may well show sufficient but distinct customer interest in each of these options, depending on levels of comfort with an electronic environment: TV services for the 'habit die-hards', PC private links for the computer-literate, managed networks for the more promiscuous consumer who might surf the Net but still be directed or channelled back to a target site. It's a question of developing the understanding and response to what target consumers need.

A few banks have already determined that a significant part of their future does lie with electronic consumers. Wells Fargo and Citibank, for example, have been particularly innovative in establishing direct customer connections. In the UK, HongKong & Shanghai's First Direct Bank decided to go after 'dissatisfied, progressive and affluent customers' and within four years had acquired over half a million customers and achieved a cost-to-income ratio some twenty percentage points below its branch average. Also in the UK, Northern Rock has been actively encouraging customers to carry out transactions by phone while consolidating the branch network – some 25 branches closed in 1997 alone. In Germany, the Bayerische Hypotheken und Wechselsbank has had success with its 'direct banking' venture. Among other things, to attract customers to switch from branch to direct telephone or PC banking, a 50% discount on bank charges was offered. At a more dedicated virtual level the USA saw the establishment in 1997 of the first banks to operate on the Internet. First Virtual and Security First are solely Internet-based financial service providers.

How far and how quickly is electronic banking going to replace traditional branch banking? An analogy can be drawn from the world of personal insurance which especially in the UK has already experienced dramatic disintermediation – even before the advent of the Internet. Direct Line, the direct telephone-based insurer, has demonstrated how easy it can be to get consumers to switch in large numbers away from traditional branch, brokers or agent distribution. Its rise has encouraged

**Figure A2.5** Personal Insurance Distribution, 1990–2005

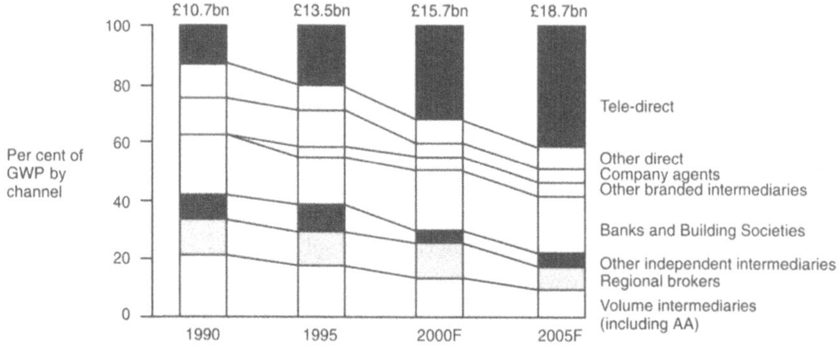

[Chart: Per cent of GWP by channel, 1990 (£10.7bn), 1995 (£13.5bn), 2000F (£15.7bn), 2005F (£18.7bn). Channels: Tele-direct; Other direct; Company agents; Other branded intermediaries; Banks and Building Societies; Other independent intermediaries; Regional brokers; Volume intermediaries (including AA).]

*Source*: ABI; Datamonitor; Mintel; CSC Kalchas.

many competitors to offer the same route to market and their total market activity is forecast to move 50% of the market direct by 2005 (Figure A2.5). The respected trade journal *Financial Adviser* predicts that:

> 'in the future being able to give the best advice and the most advantageous products to clients may well depend on an adviser's ability to deal direct and trade electronically in some form . . . the growth already seen in direct transactions is just the tip of the iceberg and providers will be well-advised to start putting their plans in place now'.

\* \* \*

Retail banking and insurance may well represent a microcosm of what will happen across the total retail environment. They may well be the 'tip of the iceberg' for the challenges and changes that will face all retail chains in the years to come. Certainly the financial services industry appears to be in the forefront. No doubt it will provide both some painful lessons but also indicative success models for dealing with the electronic revolution. Some pioneers already see branch networks as redundant, others believe they will have a role to play but in a different form. Whatever the eventual outcome, success can come only from confronting the challenge in the manner of a Citibank, a Direct Line or a First Direct and determining a clear strategic response.

# List of Figures, Tables and Plates

**Figures**

| | | |
|---|---|---|
| 3.1 | Projected food and drink sales on the Internet, 1996–2000 | 43 |
| 3.2 | Electronic shopping: the virtuous circle | 46 |
| 4.1 | Total Internet spend, 1996–2000 | 53 |
| 4.2 | Average IT spend per worker per year | 56 |
| 4.3 | Predicting when ES will reach critical mass | 66 |
| 4.4 | Growth in electronic shopping: key stages of growth, 1993–2007 | 67 |
| 5.1 | The 'ES Test' | 72 |
| 5.2 | Physical vs. virtual | 74 |
| 5.3 | Product characteristics: responsiveness to ES | 74 |
| 5.4 | Product characteristics: adding a 'sixth' sense | 76 |
| 5.5 | Familiarity/confidence check | 77 |
| 5.6 | The 'Triple I' model | 81 |
| 5.7 | Consumer categories: ES responsiveness | 82 |
| 5.8 | ES segmentation | 84 |
| 5.9 | Product characteristics' scoring | 85 |
| 5.10 | Familiarity and confidence scoring | 85 |
| 5.11 | Consumer attributes' scoring | 86 |
| 6.1 | Change in business systems | 95 |
| 6.2 | Value networks | 97 |
| 6.3 | Share of retail sales, by location (UK) | 99 |
| 6.4 | The shopping environment 2005 and beyond | 101 |
| 7.1 | What path to take? | 104 |
| 7.2 | National Westminster Bank: multichannel operator | 110 |
| 7.3 | US flagship and on-line 'mixed system' | 111 |
| 7.4 | The 'best of both' shopping experience | 115 |
| 7.5 | Comparing the options | 118 |
| 7.6 | Choosing from the ten strategic options | 119 |
| 8.1 | Plan of the store of the future | 126 |
| 9.1 | The consumer and technology | 132 |
| 9.2 | Consumer ES interest | 136 |
| 9.3 | Readiness to shop electronically | 137 |
| 9.4 | Electronic convergence | 141 |
| 10.1 | The changing commercial world | 153 |

| 11.1 | Demand, supply and enabling infrastructure | 161 |
| 12.1 | Manufacturers at a crossroads | 174 |
| 12.2 | Retailer–manufacturer relationships | 175 |
| 12.3 | Private-label grocery penetration | 176 |
| 13.1 | Picking from the ten options | 186 |
| 13.2 | Sharing responsibility | 187 |
| 13.3 | Manufacturers' strategic options comparison | 201 |
| 13.4 | Strategic choices from the ten identified | 202 |
| 14.1 | Electronic marketing evolution | 206 |
| 14.2 | The information and knowledge master | 211 |
| 14.3 | Marketing solutions framework | 211 |
| 15.1 | Key elements of effective strategy-making | 236 |
| 15.2 | Strategy-making – caught in a vicious circle | 237 |
| 15.3 | It is companies that do plan their way through which emerge as winners | 238 |
| 15.4 | The Sharing Plan | 239 |
| A1.1 | Home Depot space expansion, 1991–6 | 248 |
| A1.2 | US retail sales | 250 |
| A2.1 | Growth and decline in branch networks | 256 |
| A2.2 | US retail banks' 'wallet share': consumers with income > $75k p.a., 1990–6 | 257 |
| A2.3 | Retail banking unit transaction costs | 257 |
| A2.4 | Retail banking: cost of maintaining a current account | 258 |
| A2.5 | Personal insurance distribution, 1990–2005 | 260 |

**Tables**

| 2.1 | Electronic shopping connections | 25 |
| 2.2 | Consumer product and services | 26 |
| 2.3 | Key underlying trends behind electronic shopping | 27 |
| 4.1 | On-line households, 1996 and 2000 | 51 |
| 4.2 | Growth of European households with PCs, 1994–2000 | 51 |
| 4.3 | Penetration of US households with computers and Internet connection, 1995 and 2000 | 52 |
| 4.4 | UK growth in Internet connections, 1997–2001 | 52 |
| 4.5 | Top ten Internet shopping categories | 54 |
| 4.6 | Internet electronic shopping, by age | 55 |
| 4.7 | Internet electronic shopping, by socioeconomic groups | 55 |
| 4.8 | Confidence in using new technologies, by age group | 57 |
| 4.9 | Proportion of people who would consider remote shopping | 57 |
| 4.10 | Comfort in conducting services by computer | 58 |
| 4.11 | Attitudes to electronic commerce | 58 |
| 4.12 | Importance of electronic commerce | 59 |
| 4.13 | Singapore Internet growth, 1995–7 | 62 |
| 4.14 | Time for technology to penetrate the mass market | 64 |
| 4.15 | Possible ES shopping profiles | 69 |
| 5.1 | Five senses: primal product appeal | 73 |

## Lists of Figures, Tables and Plates

| | | |
|---|---|---|
| 5.2 | Product appeal for individual grocery product lines | 75 |
| 5.3 | Top ten brands | 78 |
| 5.4 | Consumer familiarity and confidence: 'how strong is the brand?' | 80 |
| 5.5 | Products with likely high ES potential-scoring | 87 |
| 7.1 | Ten strategic options | 105 |
| 9.1 | 1998 satisfaction index, top ten companies | 134 |
| 9.2 | Industry sector: customer satisfaction scores | 134 |
| 9.3 | Levels of shopping enjoyment | 138 |
| 9.4 | Impact of 15% of consumers buying electronically on store profitability | 140 |
| 9.5 | Types of connection possible to the Internet | 145 |
| 10.1 | March of the microprocessor, 1971–2010 | 154 |
| 10.2 | Changes in retail P&L | 157 |
| 10.3 | Top ten US brands | 158 |
| 12.1 | 'The Manufacturer's Test' | 183 |
| 13.1 | Manufacturers' ten strategic options | 185 |
| 14.1 | Seven new marketing challenges for an electronic age | 207 |
| 14.2 | Internet penetration of specific consumer segments | 219 |
| A1.1 | Four year stores plan for Thorntons, 1998–2001 | 247 |
| A1.2 | UK major new shopping centres, 1997–9 | 248 |
| A1.3 | Trends in consumer spending, 1975–95 | 252 |

**Plates**

Sainsbury's Self-service
Shopping in the electronic age
Iceland's home delivery service
Ordering your Tesco shopping on the Internet
Direct Line – insurance over the phone
Lakeside Shopping Centre and Retail Park
Web TV
The Information Skyway
Ordering from Dell over the Internet
On-line banking via Quicken at Citibank
AsiaOne Commerce
Karstadt's My-World
Dixon's website
Abbey National website

# Bibliography

**Books**

Collins, J.C. and Porras, J.I.  *Built to Last* (Century Random House, 1994).
Cronin, Mary J.  *The Internet Strategy Handbook* (Harvard Business School Press, 1996).
Davis, Stan and Davidson, Bill  *2020 Vision: Transform your Business today to succeed in Tomorrow's Economy* (Simon & Schuster, 1992).
Dertouzos, Michael L.  *What Will Be: How the new world of information will change our lives* (Harper Edge, 1997).
Drucker, Peter  *Post Capitalist Society* (Butterworth–Heinemann, 1993).
Drucker, Peter  *The New Realities in Government and Politics, Economics and Business* (Harper & Row, 1989).
Gates, Bill  *The Road Ahead* (Viking Penguin, 1995).
Hagel, John and Armstrong, Arthur  *Net Gain* (Harvard Business School Press, 1997).
Hamel, Gary and Prahalad, C.K.  *Competing for the Future* (Harvard Business School Press, 1994).
Handy, Charles  *Age of Unreason* (Random House, 1990).
Kalakota, Ravi and Whinston, Andrew  *Frontiers of Electronic Commerce* (Addison Wesley, 1996).
Komenar, Margo  *Electronic Marketing* (John Wiley, 1997).
Lynch, Daniel and Lundquist, Leslie  *Digital Money* (John Wiley, 1996).
Markham, Julian  *The Future of Shopping* (Macmillan Press, 1998).
Martin, James  *Cybercorp* (Amacom, 1996).
Morrison, Ian  *The Second Curve* (Nicholas Brealey, 1996).
Nalebuff, Barry and Brandenburger, Adam  *Cooperation* (HarperCollins, 1996).
Negroponte, Nicholas  *Being Digital* (Random House, 1995).
Petersen, Robert  *Electronic Marketing and the Consumer* (Sage Publications, 1997).
Reichheld, Frederick F.  *The Loyalty Effect* (Harvard Business School Press, 1990).
Ridpath, Michael  *Trading Reality* (William Heinemann, 1996).
Tapscott, Don  *Digital Economy* (McGraw-Hill, 1995).
Tapscott, Don  *Growing up Digital* (McGraw-Hill, 1998).
Tenner, Edward  *Why Things Bite Back* (Fourth Estate, 1996).
Wileman, Andrew and Jary, Michael  *Retail Power Plays* (Macmillan Press, 1997).
Williams, Bridget  *The Best Butter in the World – a history of Sainsbury's* (Ebury Press, 1994).

# Bibliography

**Research/Reports/Articles**

*Advertising Age* 'Web Users' (1997).
AK2A Media 'Selling on-line' (1997).
Andersen Consulting 'Understanding Consumer Direct' (1998).
Bank of America 'Home Banking Review' (1996).
Binary Compass 'Shoppers Spending Patterns On-line' (1997).
Bloomberg Business 'Internet Forecasts' (1997).
Bossard Consultants 'Home Shopping' (August 1997).
Boston Consulting Group 'Knowing your Customer, how Information will Revolutionise Food Retailing' (Coca-Cola Research Group, 1997).
British Producers and Brand Owners 'Brand names' (1997).
BT 'The Future of e-commerce' (1997).
Bureau of Transport and Communications Economics 'Home Shopping impact in the metropolitan areas', *Australian Finance Review* (1997).
*Business Week* '1998 Prognosis on Food Industry' (January 1998).
*Business Week* 'E-Shop till you Drop' (February 1998).
*Business Week* 'From Computers to Croissants' (February 1997).
*Business Week* 'Internet Communities' (May 1997).
*Business Week* 'Retailing will never be the same' (July 1993).
*Business Week* 'The Digital Frontier' (June 1997).
*Business Week* 'The Silicon Age – it's just dawning' (December 1996).
*Business Week* 'The Virtual Mall gets Real' (January 1998).
*Business Week* 'Wal-Mart Spoken Here' (June 1997).
Cable TV and R&D consortium 'Industry Wide Standards and Innovation' (July 1997).
Cap Gemini 'Internet Shopping' (January 1998).
CBI and Hewlett Packard 'Electronic Commerce' (1997).
*Central Europe Business* 'Europe Optimistic' (1997).
Centre for Financial Innovation 'The Internet and Financial Services' (1997).
*Chain Store Age* 'PC Ownership' (1996).
Chesbrough, Henry and Teece, David 'When is Virtual Virtuous?', *Harvard Business Review* (January–February 1996).
CommerceNet/Nielsen 'Shop until you drop' (1997).
*Computer Weekly* 'What's in Store?' (1996).
*Computing* 'Dawn of a New Age' (January 1998).
Coopers & Lybrand 'On-line Connections' (1997).
Corporate Intelligence on Retailing 'Retail and Leisure' (1997).
Corporate Intelligence on Retailing 'Retail Handbook' (1997).
Cowles/Simba 'Electronic Information Reports' (1997–8).
CSC 'Electronic Commerce – the next Business Frontier' (1996).
CSC 'Forming a Digital Business Strategy' (1997).
CSC 'Future of E-commerce' (1997).
CSC 'The 21st Century CEO' (1997).
*Daily Variety* 'Home Shopping in Germany' (September 1997).
Datamonitor 'Consumer on-line Shopping Revenue Opportunities' (1997).
Datamonitor 'Retail Banking' (1997).
Dataquest 'Transaction Security' (1996).

Drucker, Peter 'The New Organisation', *Harvard Business Review* (January–February 1988).
Drucker, Peter 'The New Society of Organisations', *Harvard Business Review*. (September–October 1992).
Drucker, Peter 'The Information Executives Truly Need', *Harvard Business Review* (January–February 1995).
Digital Video Broadcasting consortium 'Open Standards Report' (September 1997).
*Economist* 'Internet Shopping – the Future Mall' (November 1997).
Economist Intelligent Unit 'Home Shopping' (1996).
EIU 'Banking towards a New Millennium' (1996).
EIU 'Vision 2010' (1997–8).
Ernst & Young 'Financial Services on the Net' (1997).
Euromonitor 'Cable and Satellite Connections' (1996).
Euromonitor 'Mail Order and Home Shopping' (1996).
*Financial Times* 'The Electronic Revolution Makes Close Links with Shoppers' (March 1998).
Financial Times Reports 'Retailing on the Internet' (1996).
Financial Times Reports 'The Future of the Store' (1997).
Forrester Research 'Internet Commerce' (1997/1998).
*Fortune* 'America's Fastest Growing Companies' (September 1997).
*Fortune* 'Asian Info Tech Explosion' (August 1997).
*Fortune* 'Dell Turns the PC World Inside-out' (September 1997).
*Fortune* 'Ten Tech Trends to Bet on' (November 1997).
*Fortune* 'The Ultimate set top box' (February 1998).
France Telecom 'Aux armes, Netoyens' (September 1997).
Freedman, Reyner and Tachtemann 'European Category Management, look before you leap', *McKinsey Quarterly*, No. 1 (1997).
FT Management 'Hallo Stranger' (February 1998).
FT Management 'How the West was Wired' (November 1997).
Gartner Group 'Interactive Marketing' (1996).
Ghosh, Shikhar (Chairman of Open Market) 'Making Business Sense of the Internet', *Harvard Business Review* (March–April 1998).
GMA Planning 'Retailing and the Infotech Explosion' (September 1995).
Goldman Sachs 'The Electronic Retailing Marketplace' (1996).
Gow, Kathleen 'Risk vs. Opportunity on the Web' (1996).
Grocery Marketing 'The New Frontier' (December 1996).
*Guardian* 'High Street Heads for Superhighway' (September 1996).
GVU (Graphic Visualisation and Usability Centre, Georgia Institute of Technology) 'Web Surveys' (1995–8).
Harrington, Lorraine and Reed, Greg 'Electronic Commerce Comes of Age', *McKinsey Quarterly*, No. 2 (1996).
Healey and Baker 'Shopping Malls vs. High Street' (1997).
Henderson Crosthwaite 'Food Retail Updates' (1997, 1998).
Henley Centre 'Telemarketing' (December 1997).
IBM 'Internet Review' (1997).
ICL Retail Systems 'Home Shopping Services' (December 1997).
IDC 'Electronic Commerce Market' (February 1998).

IDC 'Web Presence' (1996).
IDC/LINK Research 'America Wired' (January 1998).
Information Strategy 'Top Companies on the Internet' (1997).
Input 'Business to Business on the Web' (1996).
*Inside Retailing* 'Operating on a Global Scale', Special Report, 3 (1997).
*Inside Retailing* 'Supermarkets Challenged to Expand Horizons', Special Report, 4 (1998).
Insight Research 'Home Shopping' (Summer 1998).
Institute for the Future 'The Future of Brands' (1994).
Institute of Grocery Distribution 'Managing Profitable Buyer/Supplier Relationships' (June 1993).
Intelliquest 'On-line Population Growth' (June 1997).
*Internet Business* 'Financial District' (April 1997).
*Internet Business* 'Surfing not Sinking, managing Internet security' (March 1997).
*Internet Information and Intelligence* 'People Hooked to the Internet' (August 1996).
*Internet World* 'Low Cost ADSL may Threaten Cable' (February 1998).
Jupiter Communications 'On-line Shopping' (1996–8).
Jupiter Communications 'On-line Kids Report' (1997).
Kalchas Tele-Direct in Insurance (1997).
Kalchas Internet growth in fmcg (1997).
KPMG 'Home Shopping' (1997).
KSA 'Electronic Shopping' (January 1998).
Kumar, Nirmalya, Professor of Marketing and Retailing, IMD 'The Power of Trust in Manufacturer–Retailer Relationships', *Harvard Business Review* (November–December 1996).
Kurt Salmon Associates 'Non Store Retailing' (Summer 1996).
Lockheed Martin Telecommunications 'Interacting Faster' (March 1997).
*Long Range Planning* 'Emerging Trends in US Retailing', Vol. 30. No. 6. (1997).
*Long Range Planning* 'The Revolution in Retailing – from market driven to market driving', Vol. 30. No. 6. (1997).
Madanmohan, Rio 'India's Long Term Vision' (August 1996).
Manning, Ric 'Manning the Wires' (1996).
Marzbani, Ramon 'Australian Internet', *Australian Financial Review* (1996).
*McKinsey Quarterly* 'The Real Impact of Internet Advertising', No. 3 (1997).
Microsoft 'Computer Penetration' (1997 report).
Mintel 'Consumer Shopping Habits – who enjoys it' (1995).
Mintel 'Leisure Shopping' (1996).
Mintel 'PC Software' (1996).
NAMNEWS 'Retail Results Digest' (1997–8).
*New York Times* 'Squeezing the Cyber Melons' (June 1997).
NFO Interactive 'On-line Shopping' (1997).
Nielsen, A.C. 'Macro Segmentation of Consumers' (1997).
NOP Research 'On-line Shopping in Britain' (1997).
NUA 'Internet Surveys' (1996–8).
Odyssey Ventures Inc. 'On-line Shopping' (March 1998).

OECD (Organisation for Economic Co-operation and Development) 'Shopping via the Internet' (February 1998).
Ovum Report 'Interactive Services' (1997).
Oxford University Institute of Retail Management 'There's Still Time' (1996).
Pira 'Corporate Web Strategies' (1997).
Price Waterhouse 'Consumer Technology Survey' (June 1997).
Rabobank International 'Pay TV' (1997).
Rayport, Jeffrey, Professor of Service Management, Harvard Business School 'Exploiting the Virtual Value Chain', *Harvard Business Review* (November–December 1995).
Reichheld, F. F. and Sasser, W. E. 'Zero Defections in Customer Service', *Harvard Business Review* (September–October 1990).
*Retail Solutions* 'Internet Briefing' (1998).
*Retail Solutions* 'Winning on the Web' (February 1998).
*Retail Verdict* 'Key Retail Stories' (1997–8).
Schneider, Fred 'Retailing on the Internet', *International Retailing Trends* (Dec. 1995).
*Singapore Business Times* 'Internet 2000' (May 1996).
Spar, Deborah, Associate Professor, Harvard Business School 'Ruling the Net', *Harvard Business Review* (May–June 1996).
*Strategy* 'Electronic Home Shopping' (July 1997).
Sun Microsystems 'Internet Essentials' (1996).
*Sunday Telegraph* 'High Street, High Noon' (March 1996).
*The Times* 'Selling by Phone' (December 1997).
Tokyo University Research Centre 'Internet Subscribers' (August 1996).
*USA Today* 'UPS jolts I-want-it-now Temperament' (August 1997).
*Verdict* 'Electronic Shopping' (1997).
*Verdict* 'Home Shopping' (1996 and 1998).
*Verdict* 'Retail Law Yearbook' (1997).
*Verdict* 'Retailing 2000' (1997).
*Wall Street Journal* 'CEO Summit on Converging Technologies' (June 1997).
*Wall Street Journal* 'Convergence' (Spring and Winter 1997).
*Wall Street Journal* 'Europe Gets Smart' (January 1998).
*Wall Street Journal* 'In Home Shopping, US Catalog Firms Jolt European Market' (December 1997).
Yankee Group 'Global Internet Shopping' (January 1998).
Zona Research 'Internet Connections' (February 1998).

# Index

ABC1 consumers 256, 257
accounting rules 157–8
ADSL (asymmetric digital subscriber lines) 144, 145
'ADSL Lite' 144
Advanced Micro Devices 241
advertising 217–19, 224
age 55
agility, organisation 233–5
Ahold supermarkets 122
Alcatel 151
alliances 140–1
Allied Domecq 41
Amana 70
Amazon.com 38–9, 79, 207, 210
America On-line (AOL) 38, 108, 171, 210
American Airlines 215
Apple 246
Argos 89–90, 113
Asda 93, 241
assets
 retailers' investment in fixed assets 246–9
 shift to knowledge 28, 153–9
ATM (asynchronous transfer mode) 143, 144
AT&T 19, 150, 151, 168
audits of web sites 224
automatic teller machines (ATMs) 2, 122–3, 254
automotive industry 40–1, 188
Avis 240
Avon 181


B&Q 106
'band together' strategy 199–200
Bank of America 92

banking 112–13
 case study 254–60
 'no-people' branches 122–3, 200
 supermarket banks 92
 *see also* financial services
banner ads 219
Barclay Square 108, 168
Barclays Bank 35, 42, 89–90, 123, 168
Barnes & Noble 38
Barnevik, Percy 224
Bartlett, Professor 20, 234
Bayerische Hypotheken und Wechselsbank 259
benefits: sharing 243
Berry, Professor 135
'best of both' strategy 113–16
 store of the future 124–8
betting shops 42
Bezos, Jeff 38
biometry 166–7
Black & Decker 179
'Blacklist of inappropriate advertisers' web home page 164
Blackwells 106–7
Blair, Tony 138
Boeing 197
Boots 92, 248
Borders 38
'brand-driven' strategy 193
brands
 customer familiarity and confidence 78–80
 values 156–8
breakeven analysis in stores 140
brewers 41
Bristol Myers Squibb 217
Britannia jeans 179
British Interactive Broadcasting 42
British Satellite Broadcasting 35

269

British Sky Broadcasting (BSB) 147, 148
British Telecom 123–4
broadband ISDN (B-ISDN) 144
'bucking the trend and revitalisation' strategy 103–4, 116–17
Burke, Raymond 10, 12
business-to-business sector 197
button ads 219

call centres 96–7
Campbell's 182
car industry 40–1, 185
Carrefour 248
cash: electronic 168–9
catalogue companies 2
CDDirect.co.uk 45
CDNow 44–5
cellular phones 150–2
CentrO 3
channel conflict 33, 35
  managing 97–8, 179–84
Chaparral Company 231
chief executives (CEOs): survey 58–9
children: kiddie power 9, 138–9
children's drop-in centre 126
CIDCO Inc. 151
Cisco Systems 197
Citibank 259
Citicorp 149
click through 223
Clinton, Bill 9, 138
clothing 75
clubs: consumer 45, 195–7
co-branded ads 220
Coca-Cola 157, 177, 178, 180
Code of Fair Information Practices 163
coffee shop: electronic 128
'collect and go' operations 41, 107–8
Collins, J.C. 19, 237–8
Commerzbank 92
communication: corporate-wide 8, 20–1, 241–2
communications technology 9
  infrastructure development 51–3
  supply-side investment 140–52
  *see also* technology

community centre 115
community-wide ES projects 46, 139
Comp USA 93
Compaq 183, 189, 246
competences: changing 94–5
complementary segmentation 180–1
CompuServe 108, 258
confidence
  consumer confidence and familiarity 15, 71–2, 77–80, 85, 87
  in using new technologies 57
consumer attributes 71–2, 80–4, 86, 87, 136–7, 259
consumer clubs 45, 195–7
consumer demand 8, 10–11, 131–40
  community ES projects 139
  customer service gap 133–6
  future growth of ES 56–9, 68–9
  impact of 15% switch 139–40
  interest in ES 136–8
  'kiddie power' and interest in ES 138–9
consumer familiarity and confidence 15, 71–2, 77–80, 85, 87
consumer trends 25–9
  eating habits 252–3
  looking to take control 27
  time poverty 27
content sites 212
control
  consumers and 27
  general regulation and 170–1
convenience 16
convenience 7–11s 100, 101
convenience shoppers ('frenzied copers') 34, 81, 82, 83, 136–7
convergence of technologies 28, 140–1, 233
Cook, Paul 242
Coopers & Lybrand 224
copper telephone wires 142–5
Coral's 42
core competences 94–5
Cormany, Doug 220
cost per click 223
cost per head 223

# Index

cost per thousand  223
costs
  banks  256, 257, 258
  retailers  156–7
Cotswold Shopping Service  44
credit cards  166, 167–8
critical mass: in the electronic
    revolution  8, 12–13, 22, 47, 63–9
Crowell, Robert J.  44
CSC  234
CU See Me  123
CUC International  210
customer satisfaction  133–6
customers
  information about for
    sellers  207–8
  information for  208–9
  managing loyalty  221–3
  understanding  213–14
  *see also* consumer attributes;
    consumer demand; consumer
    trends
Cyberway  62

Daewoo  123
Dalgety  191
Data Encryption Standard
    (DES)  165–6
database marketing  207–8
De Walt  179
decoder boxes  147–8, 148–9
delivery  *see* home delivery
Dell  36–7, 210, 229, 246
demand  *see* consumer demand
design of web site  214–15
destination specialists  100, 101
Deutsche Bank  92
Dewhirst  191
digital cash (digi cash)  168–9
Digital Equipment
    Corporation  194–5
digital television  147–8, 258
digitisation  28, 228–9
Direct Line  32–3, 109, 199, 259–60
disintermediation  232–3, 254–5
Disney  138–9
Dixons  108
Doc Marten's  4

Dockers  179
Drucker, Peter  155
Duel, Charles  63
DVB standard  147

E*Trade Group  217–18
eating habits  252–3
Edah  222
Eddie Bauer  109
Edison, Thomas  63
Ekornes  189
electricity grid  152
electronic cash  168–9
electronic coffee shop  128
Electronic Communications Privacy
    Act 1986  164
electronic information-based
    options  185–6, 193–7
electronic shopping (ES)
  critical mass  8, 12–13, 22, 47, 63–9
  definition  25
  impact on shopping and
    retailing  8, 10
  revolution is going to happen  8,
    8–9
  scenarios  47–9
  stages of growth  66–9
employee ownership  243
empowerment  238–44
encryption  165–6
entertainment
  retail/leisure parks  2–3, 99–100,
    101, 116–17
  on web site  215–16
entrants: new  44–5, 113
Ernst & Young  224
Eroski  249
'ES Test'  8, 15–16, 22–3, 71–87, 119,
    186, 235
  consumer attributes  80–4
  familiarity and confidence  77–80
  product characteristics  73–6
  scoring  84–7
ethical shoppers  82, 83
Excedrin  217
Excite Inc.  38, 218
'experimenters'  82–3, 136–7
'export' strategy  106–7

familiarity and confidence  71–2, 77–80, 85, 87
fast-packet switching  143
FBI  170–1
Fedex  96
fibre optic cable  145–7
financial services  2, 14–15, 48–9
  banking  *see* banking
  insurance  259–60
  new entrants  44
  product characteristics  76
  separate business strategy  199
Finland  46
Firefly  208
firewalls  167
First Direct Bank  34–5, 259
First Virtual  259
First Virtual Corporation  230
fixed assets  *see* assets
flagship stores and on-line delivery  110–12
Flanagans  44
flexibility  233–5
Food Ferry  44
'foods from around the world' counter  127
'forming a club' strategy  195–7
France  56, 59–60, 170, 178
Frank's Nursery  123
Fred Meyer  213–14
'frenzied copers'  34, 81, 82, 83, 136–7
fresh foods  127
future growth  47–70
  consumer acceptance  56–9
  country snapshots  59–63
  critical mass  63–9
  infrastructure development  51–3
  scenarios  47–9
  size of the Internet economy  53–4
  stages of growth  66–9
  users  54–6
  virtual store  49–50
Future Wave Software  142

Gap  215
Gates, Bill  10, 149, 250, 254
GE  150, 181, 197, 241
General Mills  215
Germany  61, 92, 164
Ghoshal, Professor  20, 234
Globalink  60
'go fully direct' strategy  200
'going with the flow'  104
Goodyear  189
Gore, Al  9
Great Universal Stores (GUS)  96–7, 109–10
grocery sector  2, 26
  change in eating habits  252–3
  product characteristics  75
  scenario  47–8
Grove, Andrew  10, 154, 217
growth: future  *see* future growth
GSM (global system for mobile communication)  150–1
Guinness  41
Gupta, Sunil  15

'habitual die-hards'  81, 82, 83
Hair, Joseph  20
hand-held terminals  121–2
Handy, Charles  21, 240
healthy eating  127
Heijn, Albert  122
Hewlett Packard  183, 230
high-street shops/malls  8, 13–14, 99–100
Hill, Kenneth  21
Hoffman-LaRoche  231–2
holidays  45
home delivery  220–1
  flagship stores and on-line delivery  110–12
  retailers and  95, 96–7
home delivery companies  42–4, 209–10
home-delivery shop  125–6
Home Depot  93, 247–8, 251
host computers  51
HOT (Home Order Television)  61
households: penetration of Internet  1, 9, 51–2
Hughes Electronics  150

## Index

IBM  167, 183, 197, 217, 258
  PC/mainframe dilemma  245–6
ICL Computers  42
immersion  81
implementation  8, 20–1, 24, 104–5
  people-empowered  238–44
industry pioneers  31–40
information
  for buyers  208–9
  local information area  127
  personal privacy  162–5
  for sellers  207–8
  sharing  241–2
Information Infrastructure Task Force  170
'information only' strategy  105–6, 193–5
information technology (IT)
  investment  56
  and marketing  233
  outsourcing  96–7
infrastructure
  consolidation  67
  development  51–3
  security issues  161–72
  supply-side investment  140–52
  *see also* technology
initial hype  66, 67
innovation  81
  retailers and  90–4
instant gratification  132–3
insurance  259–60
intangible assets  153–9
intellect  76
*Intellectual Law*  216–17
interactive kiosks  123–4, 259
interactivity  216–17
intermediaries: new  42–4, 209–10
Internet  7, 25, 50
  consumer acceptance  56–9
  country snapshots  59–63
  critical mass  63–9
  infrastructure development  51–3
  'kiddie power'  138–9
  Net Generation  9
  penetration of consumer segments  218–19
  penetration into households  1, 9, 51–2
  retail banks  258
  security issues  23, 161–72
  size of the Internet economy  53–4
  supply-side investment  140–52
  types of connection  145, 146–7
  users  54–6
  *see also* web sites
'interstitial' advertising  218
intuition  81
investment
  in IT  56
  in real estate  247–9
  S-curve analysis  66–8
  supply-side  140–52
iPhone  151
ISDN (integrated services digital network)  144, 145

Janus Software  167
Japan  177, 243
Jewel/Osco  92
John Lewis Partnership  243
Johnson & Johnson  183
Jospin, Lionel  60, 138

Ka-band  149
Keep, William  63
keyword ads  220
'kiddie power'  138–9
Kniss, Liz  46
knowledge: shift from assets to  28, 153–9
knowledge capture  231–2
Kodak  180
Kohl, Helmut  93, 138
Komenar, Margo  19
Koogle, Tim  224
Kraft  212
Kroger  92
Kumar, Nirmalya  17–18, 187

Ladbroke's  42
large shopping malls/centres  2–3, 99–100, 101, 116–17, 248
'learning, experimenting, investing' stage  67

leisure/retail parks   2–3, 99–100, 101, 116–17
Levi Strauss   4, 110, 183, 212
   bonus   243
   pioneer in ES   37
   product segmentation   179
Liedl & Schwartz   249
Lincoln Electric   241
Littlewoods   96–7
Lloyds TSB   35, 42, 258
local information area   127
Lockheed Martin   150
Logan, David   230
London: Internet pilot   46
long-term strategy   8, 19–20, 235–8
Loral Aerospace   149
L'Oreal   212
Lowe's   251
loyalty: managing   221–3
loyalty schemes   222
Lucky stores   92

Macmillan, Harold   91
Malaysia   146
managed networks   258
manufacturers   4–5, 8, 17–18, 24, 69, 173–84
   complementary segmentation   180–1
   managing channel conflict   181–4
   pioneers in ES   40–1
   product segmentation   179
   relationships with retailers   173–8
   strategic options   185–203
   territorial segmentation   180
'Manufacturer's Test'   182–3, 186, 235
market position   182–3
marketing   8, 18–19, 24, 95, 205–25
   advertising   217–19, 224
   challenges   207–10
   delivering on promises   220–1
   design of web site   214–15
   entertainment   215–16
   interactivity   216–17
   with the IT department   233
   managing loyalty   221–3
   measuring   223–4

networking the Net   219–20
overall electronic objectives   211–13
understanding customers   213–14
marketing services   96–7

Marks & Spencer   90, 109
   partnerships   188
   private label   91, 191
   retail space   247
mass marketing   67, 68
Mayo Clinic   143
McCann, Jim   34
McCaw, Craig   150
McDonald's   191, 240, 242
MCI   141
measurement systems   223–4
'mercenaries'   81, 82, 83, 136–7
Merck   192
message security   165–7
Metro Centre   3
micropayments   169
microprocessors   154
Microsoft   217, 224
   communication technology   147, 148, 149
   security   165, 168
Microsoft Network   171
microwave oven   69–70
Miller Brewing   220
Milliken, Robert   64
Minitel   59–60
'mixed system' strategy   110–12
mobile commerce   150–2
mobilisation of workforce   24, 238–44
mortgages   48–9
mother and baby clubs   45, 197
Motley Fool   196
Motorola   149
'move to private label' strategy   190–1
multimedia kiosks   123–4, 259
music sales   44–5

Nabisco   212
National Computer Board (NCB)   62

National Computer Security
    Association  167
National Information Network
    (NIN)  62
National Westminster Bank  110
Nations Bank  92
Nationwide  113
Nationwide Insurance  180
Net Generation  9
Net Nanny  164
Netscape  165, 168, 219
networking the Net  219–20
networks
    managed  258
    value networking  97, 230–1, 232
new products centre  127–8
'new wave of technology and
    equipment' stage  67
Next  61, 93, 109
Nielsen, A.C.  81, 136–7
Nike  4, 180–1, 241
Nissan  188
Nokia 9000 Communicator  151
Nordstrom  133
Norman, Archie  241
Nortel  152
Northern Foods  96
Northern Rock  259
Northern Telecom  151
Norweb Communications  152
Novon  177
number of hits  223
number of visits  223

objectives: overall electronic  211–13
'obstinate' shoppers  137–8
OECD (Organisation for Economic
    Co-operation and
    Development)  163
Olson, Ken  64
on-line communities  171
on-line delivery and flagship
    stores  110–12
on-line shopping malls  108
1–800–FLOWERS  33–4, 210
one-stop shops  91–2, 99
    virtual sites  199–200
open profiling standard  165

'order and collect' operations  41,
    107–8
organisation agility  233–5
organisation impacts  227–35
    convergence  233
    digitisation  228–9
    disintermediation  232–3
    knowledge capture  231–2
    networking  230–1
    virtualisation  229–30
Otto Versand  61
out-of-town centres  99, 248
outsourcing  96–7
'own direct' strategic options  185–6,
    197–200


Pacific  62
packaging  75
Palo Alto, California  46
parking  125
Parkinson, Thomas  43
'partnership-based' options  185–6,
    187–91
Patterson, John  231
payment security  167–70
PC banking  255
PC private dial-up services  256–8
PCs (personal computers)  148
    household penetration  51–2, 62
    IBM's dilemma  245–6
    portable  122
Peapod  43
Pentium Pro Chip  154
people-empowered
    implementation  238–44
Peripheral Vision  166
personal insurance  259–60
personal privacy  162–5
personalised shopping list  107–8
pet food suppliers  181
PetsMart  181
Petrillo, John  19
pharmaceutical companies  181
Philip Morris  182, 189
Philips  148–9, 151
Phillips, Fred  17
picking area  128

pioneers 4, 22, 31–46, 104
  consumer clubs 45
  general trial and
    experimentation 40–5
  industry pioneers 31–40
  manufacturers 40–1
  new entrants 44–5
  new intermediaries 42–4
  retailers 41–2
Pittman, Robert 54, 210
planning 235–8
Pocket Net Phone 151
Porras, J.I. 19, 237–8
portable PCs 122
Porter, Michael 234
Pressac 188
Price Waterhouse 224
privacy, personal 162–5
private label 91, 176, 177, 190–1
Procter & Gamble 183, 192, 207, 215, 224
  consumer clubs 196–7
  relationship with Wal-Mart 187–8
product characteristics 15, 71–2, 73–6, 85, 87
product segmentation 179
products 25–6, 54
profit and loss account (P&L):
  physical vs virtual 156–7
profit margins 8, 11–12, 139–40, 250
Progressive Networks 143
promotional sites 212
Publix 92
'pull-based' options 185–6, 191–3
'pursue on all fronts' strategy 109–10

Quantum 179
Quelle 61
QVC (Quality Value Convenience) 2, 35–6

Radio Shack 93
Raychem 242
ready meals 127
real estate 246–9
reception area 125
Reed, John 254
regulation 170–1

Reichheld, F.F. 222
relationship-management skills 95
'reluctant' shoppers 137–8
resisting the trend 103–4
retail banking *see* banking
retail/leisure parks 2–3, 99–100, 101, 116–17
retail margins 8, 11–12, 139–40, 250
retail price maintenance 174–6
retailers 3–4, 8, 14–15, 23, 69, 89–101
  ability to respond 90–8; changing skills and competences 94–5; making the transition 95–8; pressure to innovate 93–4
  future face of shopping 98–101
  intangible assets 156–8
  pioneers in ES 41–2
  profitability of shops
    dilemma 245–53
  relationships with
    manufacturers 173–8
  sales densities have plateaued 250
  saturation in number of shops 251
  store of the future 23, 114–16, 121–9
  strategic options 8, 16–17, 23, 103–19
  supply-chain efficiencies already captured 251–2
revenue forecasts 53–4
'revitalisation and bucking the trend' strategy 103–4, 116–17
RF (radio frequency) systems 121–2
Ridpath, Michael 49–50
Roddick, Anita 240
Rover Cars 40–1
Royal Bank of Scotland (RBS) 109

S-curve analysis 66–9
Saatchi, Maurice 13
Saatchi & Saatchi 218, 219
Safeway 107–8, 122
Sainsbury 4, 89–90, 90–1, 107, 177, 222
  pioneering projects 41–2
sales densities 250
Sara Lee 189
Sasser, W.E. 222

## Index

satellite communication  149–50
satisfaction, customer  133–6
saturation: in number of shops  251
Screen Phone  151
Seagate  197
Sears  93, 96, 181
Sears Home Services  221
Sears Roebuck  108
security  23, 161–72
  general regulation and control  170–1
  message security  165–7
  payment security  167–70
  personal privacy  162–5
Security First  259
segmentation  97–8, 179–84
  complementary  180–1
  product  179
  territorial  180
segmentation grid  84
self-help  164, 168
self-service  90–1
sellers: information for  207–8
SeniorNet  196
senses  73–6
'separate business' strategy  108–9, 199
service gap  133–5
  need to address  135–6
services  25–6, 54
services arcade  127
SET (Secure Electronic Transactions)  167–8
'set up as a separate business' strategy  108–9, 199
Sharing Plan  239–43
Sherwin Williams  189
Sheth, Jagdish  9, 12, 65
shoes  48
Shoplink  43–4
shopping centres/malls
  high-street  8, 13–14, 99–100
  large centres  2–3, 99–100, 101, 116–17, 248
  on-line  108
  out-of-town  99, 248
shops  *see* stores
Shreeve, Gavin  255

Siemens  169
sight: ES test  73–6
Silver, Jeremy  19
Singapore  61–3
Singapore ONE  62–3
'singles' nights'  82
Singnet  62
Skiing in Vail  45
skills
  changing  94–5
  sharing  242–3
Slates  179
smart cards  169–70
  loyalty system  222
'Smart Valley' project  46
SmartGATE  169
smell  73–6
Smith Barney  151
social shoppers  82
socioeconomic groups  55
Solectron  241
solutions integrator  97
Sony  148–9, 192, 198, 212
sound  73–6
space: retail  247–9
sponsorship  220
SSL (Secure Socket Layer)  168
Stern, Louis W.  13, 65
store of the future  23, 114–16, 121–9
  'best of both' illustration  124–8
  in-store technologies  121–4
store operations  95
store traffic: impact of 15% switch  8, 11–12, 139–40
stores
  danger of dying out  8, 13–14
  dilemma of profitability  245–53
  saturation in number of  251
strategic options
  manufacturers  185–203; band together  199–200; brand-driven  193; comparison  201, 202; form a club  195–7; go fully direct  200; information only  193–5; private label  190–1; separate business  199;

strategic options: manufacturers *cont.*
    strategic choices 202–3; technology-led 191–2; treat as another channel 197–8; woo the retailer 187–90
    retailers 8, 16–17, 23, 103–19; best of both 113–16; comparison and evaluation 117–19; export 106–7; information only 105–6; mixed system 110–12; pursue on all fronts 109–10; revitalise and buck the trend 103–4, 116–17; separate business 108–9; subsume into existing business 107–8; switch fully 112–13; treat as another channel 108
strategy
  clear long-term 8, 19–20, 235–8
  sharing 240
Streamline 44
structural difficulties *see* security
'subsume into existing business' strategy 107–8
Sun Trust 92
'super retailers' 176–7
super smart card 169–70
supermarket banks 92
supermarkets *see* grocery sector
supply-chain efficiencies 91, 251–2
supply-side investment 140–52
'switch fully' strategy 112–13


Taco Bell 93
Tapscott, Don 9, 17, 228
taste 73–6
technology 9, 23, 131–2
  confidence in using new technologies 57
  convergence of technologies 28, 140–1, 233
  infrastructure development 51–3
  innovations wrongly dismissed 63–4
  store of the future 121–4

supply-side investment 140–52
time to penetrate mass market 64
'technology-led' strategy 191–2
Teledisc 149–50
telephone banking 255
telephone booths 123–4
telephony
  copper telephone wires 142–5
  fibre optic cable 145–7
  wireless cellular phones 150–2
television (TV)
  digital 147–8, 258
  TV-based banking services 258
  TV cable 145–6
  web-TV set-top box 148–9
Teo Chee Hean 63
territorial segmentation 180
Tesco 4, 90, 108, 109, 122
  Extra stores 92
  pioneering projects 39–40
Third Age web site 220
Thorntons 247
3-D 49
Tide Laundry 215
Tillman, Robert 253
time poverty 27
touch 73–6
tourist centres 100, 101
Tower Records 111–12
Toyota 192, 218, 240, 242
Toys R Us 108, 199, 220
training 242–3
transaction sites 212
travel 45
'treat as another channel' strategy 108, 197–8
'triple DES' 166
'Triple I' model 81

understanding customers 213–14
Ungermann, Ralph 230
Unilever 40, 177, 182, 184, 192, 198, 232
unit transaction costs 256, 257
United Biscuits 191
United Kingdom 52, 56, 158, 163–4
  major new shopping centres 248

## Index

United Parcel Services (UPS)   20, 96, 132–3
United States (US)   92, 213, 243
  Defense Department   143
  Electronic Communications Privacy Act   164
  IT investment   56
  regulation   170
  top ten brands   157–8
*USA Today*   132–3
users   54–6, 68–9
UUNET   141

valuation rules   157–8
value networking   97, 230–1, 232
value shoppers ('mercenaries')   81, 82, 83, 136–7
values: sharing   240–1
Vavricek, Eleanor   70
vending machines   180
Victoria Wine   108
Virgin Megastore   112
virtualisation   10, 49–50, 155–6, 229–30

Waitrose   42
Wal-Mart   90, 187–8, 247, 250
Warner, Harvey   64
Watson, Thomas J.   64

web sites   18–19, 51, 89–90
  banks and   258
  designing   214–15
  directions to   219–20
  entertainment   215–16
  information-only   106, 194–5
  interactivity   216–17
  measurement systems   223–4
  on-line shopping malls   108
  role   211–13
  *see also* marketing
web TV set-top box   148–9
Wehling, Bob   18, 224, 225
Weitz, Barton A.   13, 65
Welch, Jack   21, 240, 241
Wells Fargo   92, 168, 259
Wertkauf   250
Windows 98   148
wireless cellular phones   150–2
wireless networks   121–2
women   54–5
'woo the retailer' strategy   187–90
workforce: mobilisation of   24, 238–44
World Trade Organisation (WTO)   170
WorldCom   141

Yahoo!   38, 220
You Rule School   215

**SPRINGER NATURE**

## GPSR Compliance

*The European Union's (EU) General Product Safety Regulation (GPSR) is a set of rules that requires consumer products to be safe and our obligations to ensure this.*

*If you have any concerns about our products, you can contact us on ProductSafety@springernature.com*

In case Publisher is established outside the EU, the EU authorized representative is:

Springer Nature Customer Service Center GmbH
Europaplatz 3
69115 Heidelberg, Germany

The manufacturer's authorised representative in the EU is Springer Nature Customer Service Centre GmbH, Europaplatz 3, 69115 Heidelberg, Germany. If you have any concerns regarding our products, please contact ProductSafety@springernature.com

Printed and bound by CPI Group (UK) Ltd, Croydon, CR0 4YY

23/03/2026

02076666-0016